Combating Bad Weather Part I: Rain Removal from Video

Synthesis Lectures on Image, Video, and Multimedia Processing

Editor
Alan C. Bovik, *University of Texas, Austin*

The Lectures on Image, Video and Multimedia Processing are intended to provide a unique and groundbreaking forum for the world's experts in the field to express their knowledge in unique and effective ways. It is our intention that the Series will contain Lectures of basic, intermediate, and advanced material depending on the topical matter and the authors' level of discourse. It is also intended that these Lectures depart from the usual dry textbook format and instead give the author the opportunity to speak more directly to the reader, and to unfold the subject matter from a more personal point of view. The success of this candid approach to technical writing will rest on our selection of exceptionally distinguished authors, who have been chosen for their noteworthy leadership in developing new ideas in image, video, and multimedia processing research, development, and education.

In terms of the subject matter for the series, there are few limitations that we will impose other than the Lectures be related to aspects of the imaging sciences that are relevant to furthering our understanding of the processes by which images, videos, and multimedia signals are formed, processed for various tasks, and perceived by human viewers. These categories are naturally quite broad, for two reasons: First, measuring, processing, and understanding perceptual signals involves broad categories of scientific inquiry, including optics, surface physics, visual psychophysics and neurophysiology, information theory, computer graphics, display and printing technology, artificial intelligence, neural networks, harmonic analysis, and so on. Secondly, the domain of application of these methods is limited only by the number of branches of science, engineering, and industry that utilize audio, visual, and other perceptual signals to convey information. We anticipate that the Lectures in this series will dramatically influence future thought on these subjects as the Twenty-First Century unfolds.

Biomedical Image Analysis: Tracking
Scott T. Acton and Nilanjan Ray
2006

Recognition of Humans and Their Activities Using Video
Rama Chellappa, Amit K. Roy-Chowdhury, and S. Kevin Zhou
2005

Combating Bad Weather Part I: Rain Removal from Video

Sudipta Mukhopadhyay and Abhishek Kumar Tripathi

ISBN: 978-3-031-01123-8   paperback
ISBN: 978-3-031-02251-7   ebook

DOI 10.1007/978-3-031-02251-7

A Publication in the Springer series
SYNTHESIS LECTURES ON IMAGE, VIDEO, AND MULTIMEDIA PROCESSING

Lecture #16
Series Editor: Alan C. Bovik, *University of Texas, Austin*
Series ISSN
Print 1559-8136   Electronic 1559-8144

Combating Bad Weather Part I: Rain Removal from Video

Sudipta Mukhopadhyay
IIT Kharagpur

Abhishek Kumar Tripathi
Uurmi Systems

SYNTHESIS LECTURES ON IMAGE, VIDEO, AND MULTIMEDIA PROCESSING #16

ABSTRACT

Current vision systems are designed to perform in normal weather condition. However, no one can escape from severe weather conditions. Bad weather reduces scene contrast and visibility, which results in degradation in the performance of various computer vision algorithms such as object tracking, segmentation and recognition. Thus, current vision systems must include some mechanisms that enable them to perform up to the mark in bad weather conditions such as rain and fog. Rain causes the spatial and temporal intensity variations in images or video frames. These intensity changes are due to the random distribution and high velocities of the raindrops. Fog causes low contrast and whiteness in the image and leads to a shift in the color. This book has studied rain and fog from the perspective of vision. The book has two main goals: 1) removal of rain from videos captured by a moving and static camera, 2) removal of the fog from images and videos captured by a moving single uncalibrated camera system.

The book begins with a literature survey. Pros and cons of the selected prior art algorithms are described, and a general framework for the development of an efficient rain removal algorithm is explored. Temporal and spatiotemporal properties of rain pixels are analyzed and using these properties, two rain removal algorithms for the videos captured by a static camera are developed. For the removal of rain, temporal and spatiotemporal algorithms require fewer numbers of consecutive frames which reduces buffer size and delay. These algorithms do not assume the shape, size and velocity of raindrops which make it robust to different rain conditions (i.e., heavy rain, light rain and moderate rain). In a practical situation, there is no ground truth available for rain video. Thus, no reference quality metric is very useful in measuring the efficacy of the rain removal algorithms. Temporal variance and spatiotemporal variance are presented in this book as no reference quality metrics.

An efficient rain removal algorithm using meteorological properties of rain is developed. The relation among the orientation of the raindrops, wind velocity and terminal velocity is established. This relation is used in the estimation of shape-based features of the raindrop. Meteorological property-based features helped to discriminate the rain and non-rain pixels.

Most of the prior art algorithms are designed for the videos captured by a static camera. The use of global motion compensation with all rain removal algorithms designed for videos captured by static camera results in better accuracy for videos captured by moving camera. Qualitative and quantitative results confirm that probabilistic temporal, spatiotemporal and meteorological algorithms outperformed other prior art algorithms in terms of the perceptual quality, buffer size, execution delay and system cost.

The work presented in this book can find wide application in entertainment industries, transportation, tracking and consumer electronics.

KEYWORDS

bad weather, rain, raindrop, probabilistic classifier, intensity waveform, outdoor vision and weather, meteorological properties, motion estimation, video enhancement

Dedicated to

our parents

Contents

Acknowledgments

We would like to thank our institute IIT Kharagpur for providing us a venue for this endeavor. We would like to extend our deepest gratitude to Prof. P. K. Biswas, Late Prof. S. Sengupta, Prof. A. S. Dhar, Prof. S. K. Ghosh, Prof. S. Chattopadhyay, Prof. G. Saha, Prof. S. Mahapatra and Prof. S. Banerjee for their valuable suggestions, feedback and constant encouragement.

We would also like to thank the members of the Computer Vision Laboratory. Here we had the wonderful opportunity to meet some great people. They made the work enriching and entertaining: Abhishek Midya, Jayashree, Rajat, Sumandeep, Somnath, Chiranjeevi, Manish, Jatindra, Ashis, Chanukya, Kundan, Rajlaxmi, Sandip, Vijay, Nishant and Sridevi. Special thanks to Kundan Kumar who is always ready to teach and preach LaTex. We would also thank Mr. Arumoy Mukhopadhyay for his ever-helping nature.

We feel a deep sense of gratitude to our parents and family for their encouragement and support. Finally, we would like to thank the *All mighty* for full support in this project.

Sudipta Mukhopadhyay and Abhishek Kumar Tripathi
November 2014

CHAPTER 1

Introduction

1.1 MOTIVATION

Visibility is necessary for driving. Good visibility is a requirement for passengers' safety. Reduction of visibility (owing to bad weather) causes accidents. Car accidents are common during bad weather (i.e., rain, snow and/or fog) or on slick roads (i.e., wet, snowy/slushy or icy pavement). Statistics cited by the U.S. department of transportation road weather management program states that between 1995 to 2005, an average of 17% of crash fatalities occurred due to bad weather conditions; this translates to an average of 7,400 people killed every year in the U.S. alone [U.S. Dept. of TFHA, 2012].

Bad weather not only makes driving difficult, it also makes various computer vision tasks more challenging, including road sign recognition (RSR) system [G. K. Siogkas and E. S. Dermatas, 2006], object detection, tracking, segmentation and recognition [K. Garg and S. K. Nayar, 2007]. These computer vision algorithms assume clear day weather conditions, i.e., light reflected from an object reaches the camera unaltered. Bad weather degrades not only the perceptual quality, but also the performance of various computer vision algorithms which use visual features to accomplish tasks. Hence, video post-processing has been identified as an important research topic to combat bad weather. Fig. 1.1 shows the adverse effects of bad weather. To design an outdoor vision system that is robust to bad weather, one needs to model their visual effects and develop algorithms to compensate for it.

Weather conditions differ mainly in the types and sizes of the particles present in the atmosphere [S. G. Narasimhan and S. K. Nayar, 2002]. A great effort has been taken into measuring the size of these particles (see Table 1.1). Based on the type of the visual effects, bad weather

Figure 1.1: Visual appearance of bad weather conditions: (a) rain, (b) fog, and (c) snow.

Table 1.1: Particle types and their sizes associated with various weather conditions

Condition	Particle type	Radius (μm)
Air	Molecule	10^{-4}
Haze	Aerosol	10^{-2}-1
Fog	Water droplet	1-10
Cloud	Water droplet	1-10
Rain	Water droplet	10^2-10^4

conditions are classified into two categories: steady (i.e., fog, mist and haze) and dynamic (i.e., rain and snow) [S. G. Narasimhan and S. K. Nayar, 2002]. In steady bad weather, constituent droplets are tiny (1–10 μm) and which steadily float in the air. The intensity produced at a pixel is due to the aggregate effect of the large numbers of the droplets within the pixel's solid angle. In comparison, the effects of dynamic weather are much more complex. In dynamic bad weather, constituent droplets are 1,000 times larger (0.1–10 mm) than those of the steady weather.

Rain is a very prominent weather phenomenon. Water evaporates from various water surfaces of the earth due to the heat of the sun and then a cloud is formed. When the cloud gets heavier in weight, it takes the form of drops of water that fall down to the surface of the earth [S. G. Narasimhan and S. K. Nayar, 2002]. Rain affects visibility by changing the light reflected from the road back to the driver's eye. A water drop acts as a lens which disperses the lights, so that much of it is reflected in different directions. When light strikes the raindrops, only a part of it passes through while the rest is scattered. The raindrop, therefore, obstructs some of the light reflected by objects.

CHAPTER 2

Analysis of Rain

The analysis of rain was started by atmospheric scientists [K. V. Beard and C. Chuang, 1987]. They have focused on the physical properties of rain, viz., distribution of size of raindrops, shape and velocity [J. S. Marshall and W. M. K. Palmer, 1948], [K. V. Beard and C. Chuang, 1987], [R. Gunn and G. D. Kinzer, 1949]. Later instruments and techniques were developed using active illumination and specialized detectors to estimate the rain properties [Schonhuber et al., 1994]. Various disciplines took interest in the properties of rain according to their application demand. For example, the communication engineers use Radio Detection And Ranging (RADAR) and Light Detection And Ranging (LIDAR) for signal transmission, and rain attenuates and corrupts those signals.

In computer graphics, commercial softwares are available for artificial rain rendering (Maya and Studio-Max). Scientists are also devoted to developing the mathematical models of rain [S. Starik and M. Werman, 2003]. From a distance raindrops look like a bright streak called a rain streak. Scientists have found that human vision cannot follow the individual rain streak, and hence if the sequence of rain in natural static background is changed it looks as natural as the original [S. Starik and M. Werman, 2003]. In a similar way if the patches of the rain mask are randomly changed in an artificial rain simulation environment, the effect is not visible [A. K. Tripathi and S. Mukhopadhyay, 2010]. In this chapter, we will study the properties of rain [K. Garg and S. K. Nayar, 2007].

2.1 SHAPE OF A RAINDROP

The raindrop undergoes rapid shape distortions as it falls due to air pressure. For a vision system, such change has a very insignificant effect and hence the raindrop is assumed to have fixed shape referred to as equilibrium shape. The equilibrium shape of smaller drops is close to spherical. As the size of the drops increases, it tends toward an oblate spheroid shape. Beard and Chuang approximated the equilibrium shape of a raindrop with a 10th order cosine function [K. V. Beard and C. Chuang, 1987],

$$r(\phi) = r_0(1 + \sum_{n=0}^{10} c_n \cos(n\phi)) \qquad (2.1)$$

where r_0 is the radius of the undistorted sphere, ϕ is the polar angle of elevation with $\phi = 0$ corresponds to the direction of the rainfall and c_n, $n = 0, 1, ...10$ are the shape coefficients as

Table 2.1: Shape coefficients $c_n \times 10^4$ for cosine distortion of different undistorted radius (r_0) of raindrop

r_0	n										
	0	1	2	3	4	5	6	7	8	9	10
1mm	-28	-30	-83	-22	-3	2	1	0	0	0	0
2mm	-131	-120	-376	-96	-4	15	5	0	-2	0	1
3mm	-282	-230	-779	-175	-21	46	11	-6	-7	0	3
4mm	-458	-335	-1211	-227	-83	89	12	-21	-13	1	8
5mm	-644	-416	-1629	-246	176	131	2	-44	-18	9	14
6mm	-840	-480	-2034	-237	297	166	-21	-72	-19	24	23

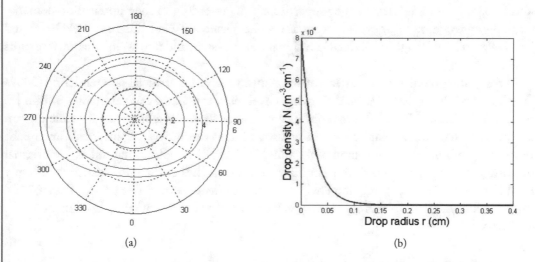

(a)   (b)

Figure 2.1: (a) Shapes of raindrops varying with diameter; (b) Raindrop size distribution with respect to the radius.

given in Table 2.1 [K. V. Beard and C. Chuang, 1987]. These coefficients and equilibrium shape of raindrops have been determined from Laplace's equation using an internal hydrostatic pressure and an external aerodynamic pressure based on measurements for a sphere, but adjusted for the effect of distortion. The shapes of the drops of varying diameter (1–5 mm) are shown in Fig. 2.1(a).

2.2   SIZE OF A RAINDROP

Raindrop density varies with the rate of rainfall as well as with size. The density of raindrops decreases exponentially with the size, and the percentage of bigger sized raindrops increases with an increase in the rate of rainfall. The most common empirical distribution of raindrop size was suggested by Marshal and Palmer [J. S. Marshall and W. M. K. Palmer, 1948]. It is commonly referred as Marshall-Palmer distribution and is given by

$$N_D = N_0 \; \exp(- \wedge D) \qquad (2.2)$$

where D is the diameter of the raindrop, $N_D \delta D$ is the number of raindrops of diameter between D and $D + \delta D$ in unit volume of space. N_0 ($0.08 \; cm^{-4}$) is the value of the N_D for $D = 0$ and $\wedge = 41 R^{-0.21} cm^{-1}$, where, R is the rainfall rate in mm/hr. Marshall-Palmer distribution of rainfall rate of $25mm/hr$ is shown in Fig. 2.1(b).

2.3   VELOCITY OF RAINDROP

As a raindrop falls, it accelerates due to the gravitational force and at the same time opposed by the drag provided by the atmosphere. The raindrops travel sufficient distance to attain a constant velocity, termed as terminal velocity. Gunn and Kinzer [R. Gunn and G. D. Kinzer, 1949] have suggested an empirical expression for the terminal velocity of a water droplet. They observed that terminal velocity v_T (cm/sec) of a water droplet of radius r (cm) could be expressed as

$$v_T = 200 \sqrt{r} \qquad (2.3)$$

In the experiment, it is noted that the terminal velocity of a raindrop depends on its diameter as bigger raindrops experience the drag to a greater extent. Foote and Toit [G. B. Foote and P. S. DuToit, 1969] developed an expression that approximates the terminal velocity (v_T (m/sec)) of the falling raindrop in terms of a polynomial of the diameter of raindrop (D (mm)). For third order polynomial approximation, the terminal velocity is given as

$$v_T = -0.19274 + 4.9625D - 0.90441D^2 + 0.056584D^3 \qquad (2.4)$$

2.4   RAIN APPEARANCE

The visual appearance of the raindrop is complex. Each raindrop can reflect and refract at the same time leading to a very complex appearance [K. Garg and S. K. Nayar, 2003]. The raindrops are randomly distributed in space and time. Due to the random distribution of raindrops, no pixel is always covered with the raindrops. Hence, there is fluctuation in pixel intensity due to rain and the same is used for synthesis and analysis of rain. The time evolution of pixel intensity in consecutive frames of a rain-affected video is shown in Fig. 2.2. The plot below shows the intensity fluctuations produced by rain at a particular pixel location.

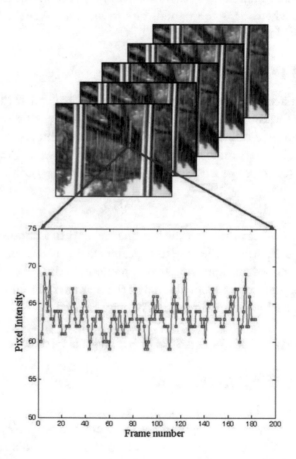

Figure 2.2: Intensity fluctuations produced by the rain at a particular pixel location.

CHAPTER 3

Dataset and Performance Metrics

3.1 RAIN DATASET

To work on the rain removal from videos first we need to get some rain videos at hand. For quantitative evaluation of rain removal algorithms, we also need the video with rain and without rain of the scene and in the same condition. As it is impossible to get them in real life, most of the tests are performed with the help of videos where the rain is simulated, and the rain mask is separately available. Some of the examples are shown in Fig. 3.1. There are a number of such sites from which such simulated video clips can be downloaded.

1. **Webpage: Photorealistic Rendering of Rain Streaks** (http://www1.cs.columbia.edu/CAVE/projects/rain_ren/rain_ren.php)

2. **Webpage: Selecting Camera Parameters for Rain Removal** (http://www1.cs.columbia.edu/CAVE/projects/camera_rain/camera_rain.php)

3. **Webpage: Detection and Removal of Rain** (http://www1.cs.columbia.edu/CAVE/projects/rain_detection/rain_detection.php)

4. **Webpage: Rain Rendering Demo** (https://sites.google.com/site/smukho/rainrenderingdemo)

5. **Webpage: Simulation of Rain in Videos** (http://www.cs.huji.ac.il/~werman/Papers/rain/rain.html)

6. **Webpage: Meteorological Approach for Detection and Removal of Rain from Videos** (http://www.ecdept.iitkgp.ernet.in/web/faculty/smukho/docs/rain_meteo/rain_meteo.html)

7. **Webpage: Rain and Snow Removal via Spatio-Temporal Frequency Analysis** (http://www.cs.cmu.edu/~pbarnum/rain/rainAndSnow.html)

3.2 PERFORMANCE METRICS

For quantitative evaluation of the performance of any rain removal algorithm performance metrics are required. In this section a few important metrics were used for the evaluation.

Figure 3.1: Video sequences frequently used for the simulation experiment with rain removal algorithms; (a) "black car01" [K. Garg and S. K. Nayar, 2006]; (b) "blue car" [S. Starik and M. Werman, 2003]; (c) "football" [A. K. Tripathi and S. Mukhopadhyay, 2010]; (d) "street01" [K. Garg and S. K. Nayar, 2007].

3.2.1   MISS, FALSE DETECTION AND ERROR

Miss detection means that a rain pixel is detected as a non-rain pixel by the rain detection algorithm. Low value of miss detection indicates good performance. False detection means that a non-rain pixel is detected as a rain pixel by the rain detection algorithm. Low value of false detection indicates good performance. Numbers of miss and false detection are calculated at arbitrary 3D window size for evaluation of performance of any rain detection algorithm. The sum of miss and false detection provides the number of erroneous pixels.

3.2.2   RAIN REMOVAL ACCURACY

Quantitative analysis is also performed in terms of the amount of the rain removed and scene distortions as rain cannot be removed without disturbing any background pixel [Barnum et al., 2007, 2008]. The equations are given for the grayscale images. For color images evaluation is carried out by combining the results after separately computing the error for each channel. Since rain only increase the brightness, a rain removal algorithm should either decrease the brightness or leave it unchanged. Removal accuracy (RA) is measured by computing the difference of the rendered rain component (True) and estimated rain component (Est.).

$$True(x, y, t) = \text{Rain rendered frame}(I(x, y, t)) - \text{Original clear frame}(C(x, y, t)) \qquad (3.1)$$

$$Est.(x, y, t) = \text{Rain rendered frame}(I(x, y, t)) - \text{Restored frame}(I_0(x, y, t)) \qquad (3.2)$$

$$RA(x, y, t) = \begin{cases} True(x, y, t) - Est.(x, y, t) & Est.(x, y, t) > 0 \\ Est.(x, y, t) & Est.(x, y, t) < 0 \end{cases} \qquad (3.3)$$

The accuracy of the rain removal algorithm is measured based on two percentage measures of amount of the rain removed (H) and amount of the scene distortion (E), given as [Barnum et al., 2007],

$$H = 100 \times \left(1 - \frac{\sum_{x,y,t} RA(x, y, t) : RA(x, y, t) > 0}{\sum_{x,y,t} True(x, y, t)} \right) \qquad (3.4)$$

$$E = 100 \times \left(\frac{\sum_{x,y,t} RA(x, y, t) : RA(x, y, t) < 0}{\sum_{x,y,t} C(x, y, t)} \right) \qquad (3.5)$$

It is interesting to note that H and E are miss and false detection as percentage of total rain pixels.

3.2.3 VARIANCE

In a video with constant background (when there is no moving object) there is no change in the intensity of the pixels in consecutive frames. Raindrops and moving objects produce a change in the intensity values in few consecutive frames. If rain is removed from the video then this intensity variation is reduced. In other words, variance contributed by rain is reduced [K. Garg and S. K. Nayar, 2005]. This remains true even when moving objects are present. Variance calculated over consecutive frames is named as temporal variance. Temporal variance (σ_t^2) of the intensities of n consecutive frames at a particular pixel position is given by

$$\sigma_t^2(x, y) = \frac{1}{n} \sum_{i=1}^{n} (I(x, y, i) - \bar{I}(x, y))^2 \tag{3.6}$$

$$\text{where} \quad \bar{I}(x, y) = \frac{1}{n} \sum_{i=1}^{n} I(x, y, i)$$

As the rain pixels are not known, a patch of pixels is selected at random in the rain region and the temporal variance is calculated. For the video the temporal variance is the average of the variance of the selected pixels, is given as:

$$\sigma_t^2 = \sum_x \sum_y \sigma_t^2(x, y) \tag{3.7}$$

This spatiotemporal variance (σ_{st}^2) is calculated over a 3D window. Spatiotemporal variance of the intensities of n consecutive frames for a window size $w \times w$ centered at a particular pixel position is given by

$$\sigma_{st}^2(x, y) = \frac{1}{w^2 n} \sum_{k=-\frac{(w-1)}{2}}^{\frac{(w-1)}{2}} \sum_{l=-\frac{(w-1)}{2}}^{\frac{(w-1)}{2}} \sum_{i=1}^{n} (I(x + k, y + l, i) - \bar{I}(x, y))^2 \tag{3.8}$$

where

$$\bar{I}(x, y) = \frac{1}{w^2 n} \sum_{k=-\frac{(w-1)}{2}}^{\frac{(w-1)}{2}} \sum_{l=-\frac{(w-1)}{2}}^{\frac{(w-1)}{2}} \sum_{i=1}^{n} I(x + k, y + l, i)$$

where $I(x, y, i)$ is the intensity of a pixel at position (x, y) in frame i. Spatiotemporal pixel intensity waveform at the same position for rain and rain-removed videos with same background is shown in Fig. 3.2. Mean of the spatiotemporal variances of the waveforms for 100 rain pixels in 15 consecutive frames of the "test" rain video are 27.92 (for 3D window $3 \times 3 \times 15$) and 43.39 (for 3D window $5 \times 5 \times 15$). Corresponding values for the rain-removed videos are 3.12 (for 3D

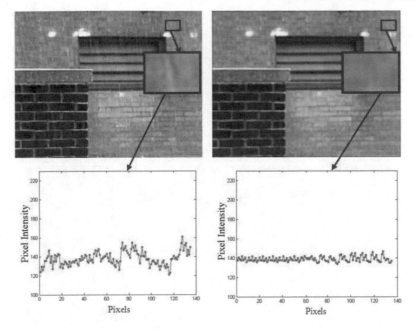

Figure 3.2: (Left) Frame from the "test" rain video and spatiotemporal intensity waveform at a pixel position; (Right) The same frame from the same rain video after rain removal by temporal rain removal technique and spatiotemporal intensity waveform at the same pixel position.

window $3 \times 3 \times 15$) and 18.38 (for 3D window $5 \times 5 \times 15$). From the results it is clear that the removal of rain from the videos reduces the spatiotemporal variance. Lower value of spatiotemporal variance indicates better performance.

The temporal and spatiotemporal blurring can also reduce the variance but it introduces blurring artifacts near the moving object boundaries. Hence, variance and spatiotemporal variance are successful as metrics as long as the rain removal technique uses discriminative inpainting. If temporal and spatial blurring is proposed for rain removal from videos, the temporal and spatiotemporal variance need to be used in conjunction with a blurring metric.

CHAPTER 4

Important Rain Detection Algorithms

4.1 FRAMEWORK

A generic framework of rain detection and removal is shown in Fig. 4.1. It is a common practice to detect the rain pixels, i.e., pixels affected by rain, for inpainting [K. A. Patwardhan and G. Sapiro, 2003, Patwardhan et al., 2005, Gu et al., 2004]. The algorithms first detect all possible rain candidates and refine these candidates to reduce the number of false candidates. The algorithms require a certain number of consecutive frames to estimate the rain-affected pixels [K. Garg and S. K. Nayar, 2004]. For the refinement some rain properties or rain models are used which vary from algorithm to algorithm. Once potential rain candidates are found, these rain pixels are inpainted. For most of the algorithms, the rain detection stage remains the same except for the number of frames used. These algorithms differ from each other in the refinement stage.

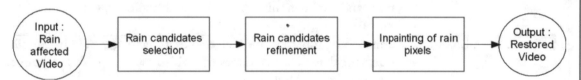

Figure 4.1: Block diagram of rain detection and removal.

4.2 SELECTED RAIN DETECTION ALGORITHMS

In the past few years, many algorithms have been proposed for the removal of rain. These algorithms can handle rain video with static background, dynamic background or both. When there are moving objects in the video, it is termed as the dynamic background video otherwise the static background video. These videos can be captured with a static camera or a moving camera. Comparison of various rain removal algorithms is shown in Table 4.1. These algorithms work either on the frequency domain or the time domain. Thus, these algorithms can be categorized in the frequency domain-based and time domain-based approaches. In this chapter, only a few important algorithms will be discussed.

Table 4.1: Comparison of various rain removal algorithms

Method	Model or properties used	Video handled
Starik et al. [2003]	Temporal properties	Static video captured with static camera
Garg and Nayar [2004]	Temporal properties for detection and photometric constraint for refinement	Static and dynamic background video captured with static camera
Garg and Nayar [2005]	Adjustment of camera parameters	Static and dynamic background video captured with static and moving camera
Zhang et al. [2006]	Chromatic properties for detection and temporal properties for refinement	Static background video captured with static camera
Barnum et al. [2007]	Blurred Gaussian model (works in frequency domain)	Static and dynamic background video captured with static and moving camera
Park et al. [2008]	Temporal properties for detection and Kalman filter for removal	Static background video captured with static camera
Brewer et al. [2008]	Temporal properties for detection and shape characteristics of rain streaks for refinement	Static and dynamic background video captured with static camera
Subhani et al. [2008]	Spatio-temporal consistency constrains for detection and refinement	Static and dynamic background video captured with static camera
Zhao et al. [2008]	Histogram model for the detection	Static background video captured with static camera
Liu et al. [2008]	Chromatic properties for detection and refinement	Static and dynamic background video captured with static camera
Liu et al. [2009]	Chromatic properties for detection and refinement	Static and dynamic background video captured with static camera
Bossu et al. [2011]	Gaussian mixture model for detection and orientation of histogram for refinement	Static and dynamic background video captured with static and moving camera

4.2.1   TIME DOMAIN-BASED APPROACH

Working in time domain results in blurring artifacts in the output rain-removed video due to a large number of false positives arising out of the moving objects. These time domain-based approaches utilized additional properties of rain to reduce the artifacts. These properties are temporal, chromatic or hybrid (i.e., combination of both) in nature.

Temporal properties-based approach

An early method for reducing rain appearance proposed by S. Starik and M. Werman [2003] used temporal median filter for each pixel. This approach was successful at reducing rain effects from the videos of the static scene, but resulted in blurring artifacts in video with a scene motion or dynamic background.

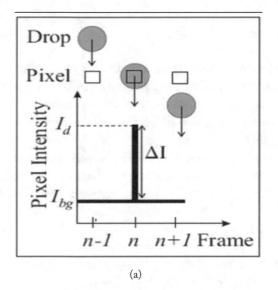 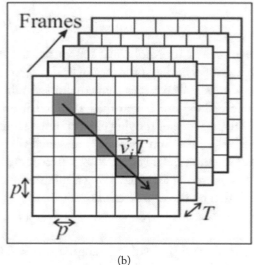

(a)   (b)

Figure 4.2: (a) Change of intensity due to rain at a pixel location; (b) spatio-temporal distribution of rain forming rain streak over time.

K. Garg and S. K. Nayar [2007, 2004] developed a comprehensive model for the visual appearance of rain. Based on this model, Garg and Nayar presented an algorithm for detection and removal of rain. The authors have observed that the rain is randomly distributed in space, and the distribution is wide sense stationary. Rain drops fall with high velocity and a raindrop occludes a pixel for very little time as given in Fig. 4.2(a). The projection of rain on the 2D image plane can be represented by a binary field given as

$$b(\vec{r}, n) = \begin{cases} 1, \text{ if drop projects to location } \vec{r} \text{ at } n \text{ th frame;} \\ 0, \text{ otherwise.} \end{cases} \qquad (4.1)$$

where $\vec{r}$ represents the spatial coordinates in the image and n is an index of the frame under consideration. Considering the rain drops in a straight line, the rain drops are spatio-temporally correlated, and form rain streaks as shown in Fig. 4.2(b). Here assumptions are that 1) the motion of background is slow, causing constant irradiance of background over the exposure time and 2) the average irradiance of raindrop can be assumed to be constant for pixels that lie on the same rain streak, because the brightness of the drop is weakly affected by the background intensity. It is assumed that a raindrop affects a particular pixel for a single frame and in very few cases in two consecutive frames. If a raindrop covers a pixel then intensity change due to rain between the pixel in current and previous frame is equal to the intensity difference between the pixel in current frame and next frame. Garg and Nayar assumed if the background remains stationary in three frames, then the intensities I_{t-1} and I_{t+1} must be equal and change in intensity ΔI due to raindrop in the tth frame must satisfy the constraint

$$\Delta I = I_t - I_{t-1} = I_t - I_{t+1} \geq c \tag{4.2}$$

where c is a threshold that represents the minimum intensity change due to rain. When the background is not stationary over the three frames, many rain pixels may be missed. However, the rain streak passing through missed rain pixels may be detected in the subsequent frames. Hence, Garg and Nayar used a sufficient number of frames (say, 30 frames) for rain detection. In the presence of moving objects in the scene, this method produces a lot of false detentions. To reject the false rain pixels it is assumed that raindrops follow the linear photometric constraint [K. Garg and S. K. Nayar, 2004]. According to the photometric constraint, changes in intensities ΔI observed at all pixels along a rain streak are linearly related to the background intensities I_{bg} occluded by the rain streak as

$$\Delta I = -\beta I_{bg} + \alpha \tag{4.3}$$

where slope β lies within a range $0 < \beta < 0.039$ and $\alpha = \tau E_d$. E_d is the irradiance due to raindrop and time τ is far less than the camera exposure time, and typical value of τ is $1.18ms$. For each streak in frame t, authors verified whether the intensity changes ΔI along the streak are linearly related to background intensities I_{t-1}. The slope β of the linear fit is estimated. Then, only the rain streaks that satisfy the photometric constraint, or whose slopes lie in the acceptable range of $\beta \in [0, 0.039]$, are accepted. In heavy rain, raindrops could affect the same position in two or more consecutive frames. The photometric model assumes that raindrops have almost the same size and fall at the same velocity. It is also assumed that pixels that lie on the same rain streak have the same irradiance because the brightness of the drop is weakly affected by the background. It is found that the variation of the size and velocity of raindrops violate the assumptions of the photometric model. This method fails to discriminate between rain pixel and moving object pixels when rain becomes heavy or light in the video or if rain is distributed over a wide range of depth. Thus, all the rain streaks do not follow the photometric constraint. Thus, it gives a lot of miss detection. This method requires 30 consecutive frames for the removal of rain.

Chromatic properties-based approach

Liu et al. [2008] proposed an algorithm for rain removal using chromatic properties in order to improve the quality of the video. These chromatic properties are similar to that developed by K. Garg and S. K. Nayar [2007, 2004]. Along with these assumptions, the fact that the change in irradiance is the same in all three (R, G, B) channels, help to build the two relations at the same pixel location in two consecutive frames: 1) relation of background pixel and rain-affected pixel, 2) relation of two rain-affected pixels. But these relations fail to discriminate between the rain and moving object pixels. Hence, Liu et al. estimated the density of pixels in the selected area using a discriminating function. The discriminating function for the separation of rain and moving object pixels is as follows

$$E_r = \frac{\int_\Omega r(x)dx}{\int_\Omega f(x)dx} > \delta_1$$

$$E_m = \frac{\int_\Omega m(x)dx}{\int_\Omega f(x)dx} < \delta_2 \qquad (4.4)$$

$$E_u = \frac{\int_\Omega (f(x) - r(x) - m(x))dx}{\int_\Omega f(x)dx} < \delta_3$$

where Ω is the area under consideration; $f(x)$ is indicator function of pixels in Ω, if $x \in \Omega$, $f(x) = 1$; $r(x)$ is indicator function of rain pixels in Ω, if x meets the detecting function, $r(x) = 1$; $m(x)$ is indicator function of moving object pixels in Ω, if x meet the detecting function, $m(x) = 1$. In fact, E_r, E_m and E_u show the density of rain pixels, moving object pixels and unaffected pixels respectively and help to determine the rain region. Values of parameters δ_1, δ_2 and δ_3 depend on the video sequence. For the removal of rain, Liu et al. have considered only the rain-affected pixels on the stationary background. Rain-affected pixels are slightly brighter than background pixels. For every two related rain pixels, Liu et al. replaced the larger value with a smaller one. To estimate the background value I_b of rain pixels, they searched the neighborhood on consecutive frames as

$$I_b = \inf \{x \mid x \in \Omega' \backslash M\} \qquad (4.5)$$

where Ω' is the search space and M is the space of moving object pixels in Ω'. This method requires 30 frames to reduce the rain visibility. This method fails to detect all possible rain pixels and gives large false pixels in case of moving objects.

Liu et al. [2009] proposed another method for the removal of rain by using chromatic-based properties in rain-affected videos. The distribution of the rain candidates is estimated to reject improperly detected pixels. Using this detecting function, the frame is divided into two parts: background and foreground. Then the removal algorithm is applied separately in these two parts. Here the Kalman filter is used to reduce the rain visibility. This method fails to detect all possible rain streaks. The reason could be that chromatic property is not satisfied in practice as described earlier. This method requires at least three consecutive or two arbitrary frames other than the current frame for the removal of rain. It makes the pixels at the same locations of backgrounds in each frame similar and shows a better perceptual image quality. This method does not use any information about the shape and velocity of raindrops. Therefore, it is effective in various rain conditions such as heavy rain, light rain and moderate rain. If rain streaks cover the same location in two arbitrary frames then this method fails to detect the rain streaks. This method fails when there are heavy rain and various moving objects in the background.

Hybrid approach

Zhang et al. [2006] proposed a method based on the chromatic and temporal properties. Temporal property suggests that a pixel is never always covered by rain in the video. Chromatic property states that changes of intensities in R, G and B color components due to the raindrops are approximately the same. In practice, these variations across the color components are bound by a small threshold. If R, G, B and R', G', B' are color components of a same pixel location in two consecutive frame then changes in the color components $\Delta R = R - R'$, $\Delta G = G - G'$ and $\Delta B = B - B'$ are approximately same. It can be represented in another way as

$$e = (\Delta R - \Delta G)^2 + (\Delta R - \Delta B)^2 + (\Delta B - \Delta G)^2 \qquad (4.6)$$

where error e is bounded by a small threshold. Rain drops do not always cover a particular pixel position in all frames according to the temporal property of rain. Hence, the intensity histogram of a pixel in the video captured by static camera exhibits two peaks. Here k-means clustering is used to identify the two peaks. These two peaks correspond to the background pixels and rain pixels. This clustering method is effective only in detecting rain from a static background when there are no moving objects. It is also found that slow moving objects also follow this chromatic property. Zhang et al. applied dilation and Gaussian blurring on the detected rain pixels and used them as the alpha (α) channel to remove rain streaks by α-blending. This step improves the rain removal result. New color C of the rain pixel is estimated by the α-blending of its rain-affected color C_r and background color C_b as

$$C = \alpha C_b + (1 - \alpha) C_r \qquad (4.7)$$

In videos with dynamic background, this method produces a degradation in the perceptual image quality. This method uses all the frames available in a sequence for the removal of the rain. The advantage of this algorithm is that it can handle both light rain and heavy rain conditions, as well as rain in focus and rain that is out of focus.

4.2.2   FREQUENCY DOMAIN-BASED APPROACH

Barnum et al. [2007, 2008] proposed a method for the detection and removal of rain streaks by using frequency information of each frame. Here a blurred Gaussian model is used to approximate the blurring produced by the raindrops, and a frequency-domain filter is used to reduce its visibility. Barnum et al. assumed that detecting individual streaks is very difficult, so they combined the streak model with statistical properties of rain to create a model for the overall effect of rain in the frequency domain. It is assumed that the particle moves in space, and the image it creates is a linear motion blurred version of the original Gaussian. If the raindrop is large or close to the camera, the rain streak will be wide. If the raindrop is falling faster, then it will be blurred into a longer streak. The equation of the blurred Gaussian model, centered at image location $\mu = [\mu_x, \mu_y]$, with orientation θ, is given by

$$g(x, y; a, z, \theta, \mu) = \int_0^{l(a,z)} \exp\left(-\frac{(x - \cos(\theta)\gamma - \mu_x)^2 + (y - \sin(\theta)\gamma - \mu_y)^2}{b(a,z)^2}\right) d\gamma \quad (4.8)$$

where a and z are the diameter of raindrop and depth from the camera; b and l are the breath and length of the rain streak. In the notation, a semicolon is used to differentiate the image location variables from others parameters. The pixel intensity due to rain in one frame at a given location (x, y) is the sum of the streaks created by all N visible drops

$$\sum_{d=1}^{N} g(x, y; a_d, z_d, \theta_d, \mu_d) \quad (4.9)$$

The same relationship is true in the frequency domain, i.e., Fourier transforms of pixel intensities is the sum of Fourier transforms of each streak g. Since magnitude of rain streaks is location independent and similar for each streak, thus

$$\sum_{d=1}^{N} ||G(u, v; a_d, z_d \theta_d, \mu_d)|| \quad (4.10)$$

Instead of trying to determine each of N streaks, authors used a model R that has frequencies proportional to main streak and scaled by overall brightness Λ.

$$R(u, v; \Lambda, \theta_{max}, \theta_{min}) = \Lambda \int_{\theta_{min}}^{\theta_{max}} \int_{a_{min}}^{a_{max}} \int_{z_{min}}^{z_{max}} z^2 ||G(u, v; a, z, \theta)|| \, dz \, da \, d\theta \quad (4.11)$$

Barnum et al. predicted this magnitude is constant in temporal frequency w. Hence

$$R(u, v, w; \Lambda, \theta_{max}, \theta_{min}) = R(u, v; \Lambda, \theta_{max}, \theta_{min}) \quad (4.12)$$

To use this model in a movie $m(x, y, t)$, only a single intensity Λ needs to be estimated per frame and only one orientation θ per sequence. Λ can be estimated by taking a ratio of the median of all frequencies, except for the constant temporal frequency $w = 0$.

$$\Lambda \approx \frac{median\left(||M(u, v, w)||\right)}{median\left(R(u, v, w, \Lambda = 1, \theta_{max}, \theta_{min})\right)} \quad (4.13)$$

where M is the Fourier transform of m. The correct θ is found by minimizing the difference between the model and the estimate

$$\arg\min_{\theta} \int \int \left(||R(u, v; \Lambda, \theta_{max}, \theta_{min}|| - \bar{R}(u, v) \right)^2 \, du \, dv \qquad (4.14)$$

where $\bar{R}$ is an estimate of the important frequencies, obtained by the standard deviation over time for each spatial frequency, for T frames

$$\bar{R}(u, v) = \sqrt{\frac{1}{T} \sum_{t=1}^{T} \left(||M(u, v, t|| - ||\bar{M}(u, v)|| \right)^2} \qquad (4.15)$$

Working in the frequency domain has some advantages as well as disadvantages. An advantage is that it allows for fast analysis of repeated patterns. Another advantage of this algorithm is that it is effective for videos with both scene and camera motions. The disadvantage is that changes made in the frequency domain do not always cause a pleasing effect in the spatial domain. Frequency-based detection has errors when frequencies corresponding to rain are too cluttered. This blurred Gaussian model is suitable when the rain streaks are prominent. Thus, this algorithm fails to detect the rain streak when it is not sharp enough.

CHAPTER 5

Probabilistic Approach for Detection and Removal of Rain

5.1 INTRODUCTION

In this chapter, novel and efficient temporal and spatiotemporal rain removal algorithms are presented. These algorithms use the time evolution properties of the pixel intensity for the separation of rain and non-rain pixels. Temporal rain removal algorithms use a large number of consecutive frames that need large buffer size and cause delay. Hence to reduce the buffer size and delay, a spatiotemporal rain removal algorithm is introduced. The temporal and spatiotemporal algorithms operate only on the intensity plane rather than on all the three color components, reducing the complexity and execution time significantly.

This chapter is organized as follows. In Section 5.2, temporal analysis of the pixel intensity is performed. Here difference between the nature of rain and non-rain pixel is observed. In Section 5.3, rain removal algorithms are presented. In Section 5.5, videos used for simulation are discussed. In Section 5.6, simulation and results are presented. Section 5.7 concludes this chapter.

5.2 TEMPORAL ANALYSIS OF PIXEL INTENSITY

In a rain video, raindrops are randomly distributed in the space. Due to the random distribution of raindrops, a pixel at a fixed location is not always covered by raindrops in each frame. This phenomenon causes the temporal fluctuation of pixel intensity. The raindrop increments the intensity value of the pixel it is covering without affecting the chrominance values [K. Garg and S. K. Nayar, 2004]. These fluctuations are very small in nature. Hence, the time evolution of pixel intensity can be an indicator of rain. Here, it should be mentioned that the pixel intensity can change in consecutive frames either due to the presence of rain or moving objects. The time evolution of pixel intensity at a particular position in the region affected by rain is quite different from the time evolution of pixel intensity due to a moving object. The intensity waveforms for rain and moving object pixels are shown in Fig. 5.1. For rain pixels, intensity values below and above the mean intensity are found to be more symmetric than the pixels belong to moving objects. In intensity waveform, range of pixel intensity (y) axis shows that the intensity variations caused by rain are small in comparison with moving objects. Thus, the range of intensity values could be a good attribute for the discrimination of rain and moving object pixels.

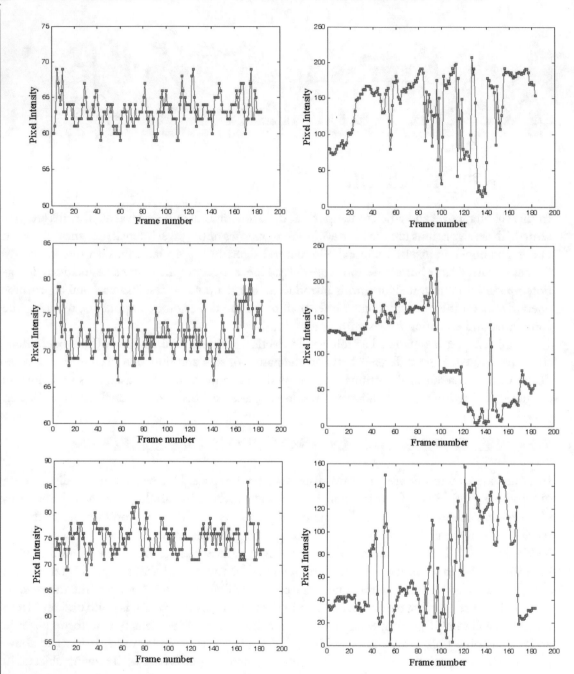

Figure 5.1: First column shows the plot of pixel intensity in consecutive frames for three pixels affected by rain. Second column shows the intensity plot for three pixels in presence of moving objects instead of rain.

The nature variation of the intensity waveform of rain-affected pixels and pixels on moving objects can also be measured by other statistics, viz., skewness. Magnitude of skewness of the data sample $x_1, x_2, x_3, ..., x_n$ can be given by eqn.(5.1).

$$K = \left| \sum_{i=1}^{n} \left(\frac{x_i - \bar{x}}{s} \right)^3 \right| \qquad (5.1)$$

where $\bar{x}$ is the sample mean and s is the sample standard deviation.

Since $(x - \bar{x})^3$ is positive when $(x - \bar{x})$ is positive and negative when $(x - \bar{x})$ is negative, data samples greater than the sample mean contribute positive terms to the sum, while data samples less than the sample mean contribute the negative terms. Perfectly symmetric data have skewness = 0, because the contributions from positive and negative terms cancel out. As the symmetry of the data decreases value of skewness increases. Table 5.1 shows the values of skewness for the temporal pixel intensity waveform shown in Fig. 5.1. In Table 5.1, rain pixels produce a

Table 5.1: Skewness of the temporal intensity waveform shown in Fig. 5.1

S.N.	Rain pixel	Non-rain moving object pixel
1	43.3231	161.0587
2	42.3491	106.8827
3	49.5094	104.0410

small value of skewness in comparison with the non-rain moving object pixels. It means intensity variations produced by the rain pixels are more symmetric than the non-rain moving object pixels.

Symmetry of the intensity variations can also be examined by the Pitman test for the symmetry [P. Sprent and N. C. Smeeton, 2001]. Temporal pixel intensity waveforms show that a rain region produces a more symmetric plot than a non-rain region. The basic idea behind the test is as follows. In a sequence, each sample will differ from the mean with some positive or negative deviation. For data with a symmetric distribution, the sums of positive and negative deviations from the mean should not differ greatly. Thus, this measure can effectively discriminate symmetric distribution of rain pixel intensity evolution and asymmetric distribution of pixel intensity evolution in non-rain region.

5.3   RAIN REMOVAL ALGORITHM[1]

In this section, we start by the selection of possible rain candidate pixels. Then we discuss their refinement for the selection of the potential rain candidate pixels. Block diagram of the probabilistic temporal rain removal algorithm is shown in Fig. 5.2.

[1]Patent application No. 1284/KOL/2010, titled as METHOD AND APPARATUS FOR DETECTION AND RE-MOVAL OF RAIN FROM VIDEOS USING TEMPORAL AND SPATIOTEMPORALPROPERTIES, published on 19/10/2012. PCT Application No. PCT/IN2011/000778 dated 11-Nov-2011.

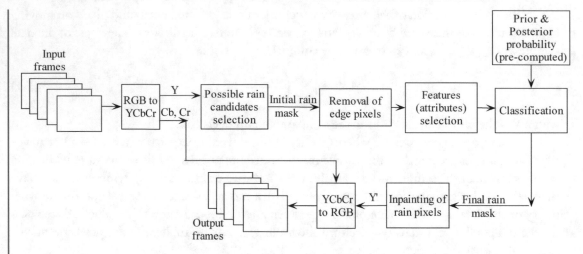

Figure 5.2: Block diagram of the probabilistic temporal rain removal algorithm.

5.3.1   DETECTION OF RAIN

Here it is considered that heavy rain may affect the pixel intensity at a particular position in one, two or maximum three consecutive frames. Hence, the intensity changes of five consecutive frames have been examined for the detection of possible rain pixels. Schematic and pictorial views of the intensity changes of the consecutive frames due to rain are shown in Fig. 5.3. To find the possible rain candidate pixels in the nth frame, we need to consider intensities $I_{n-2}, I_{n-1}, I_n, I_{n+1}$ and I_{n+2} at each pixel location corresponding to 5 consecutive frames $n-2, n-1, n, n+1$ and $n+2$, respectively. Change in intensity ΔI must satisfy the following constraints.

Constraint, when rain corruption limited to one frame,

$$(I_n - I_{n-1}) \geq t \ \& \ (I_n - I_{n+1}) \geq t \tag{5.2}$$

Constraints, when rain corruption limited to two consecutive frames,

$$(I_n - I_{n-1}) \geq t \ \& \ (I_{n+1} - I_{n+2}) \geq t \ \& \ |(I_n - I_{n+1})| \leq t_1$$
$$\text{or} \tag{5.3}$$
$$(I_{n-1} - I_{n-2}) \geq t \ \& \ (I_n - I_{n+1}) \geq t \ \& \ |(I_{n-1} - I_n)| \leq t_1$$

Constraint, when rain corruption limited to three consecutive frames,

$$(I_{n-1} - I_{n-2}) \geq t \ \& \ (I_{n+1} - I_{n+2}) \geq t \ \& \ |(I_{n-1} - I_n)| \leq t_1 \ \& \ |(I_n - I_{n+1})| \leq t_1 \tag{5.4}$$

where t is a threshold that represents the minimum change in intensity due to a raindrop, and $t_1 < t$ (t and t_1 are small positive integer). Due to the presence of moving objects, this detection

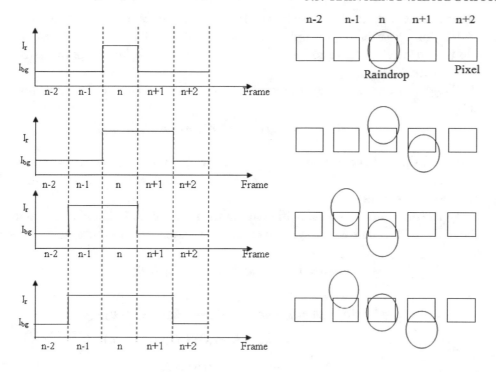

Figure 5.3: (a) Schematic and (b) pictorial views of the intensity changes of a pixel in the consecutive frames due to rain.

process contains some false rain pixels candidates. Inpainting [K. A. Patwardhan and G. Sapiro, 2003, Patwardhan et al., 2005, Gu et al., 2004] of these false rain pixels produces some unrealistic effects. These effects are more visible on the edges. Thus, this detection process requires some refinement. For the refinement of rain pixels, first moving and static edge pixels [Xu et al., 2005] of the current frame are removed. This elimination of edge pixels also helps in producing more pleasing result. Edge information can be obtained according to the eqn. (5.5).

$$E_i(p) = \begin{cases} No\ edge & \text{if } CE_i(p) = 0 \\ Static\ edge & \text{if } CE_i(p) = CE_{i-1}(p) = CE_{i+1}(p) = 255 \\ Moving\ edge & \text{otherwise} \end{cases} \qquad (5.5)$$

where $E_i(p)$ is the edge information of the pth pixel of frame i and CE is the Canny edge operator. Here Canny edge operator is selected because of its merits [J. F. Canny, 1986]. This elimination of edge pixels prevents the blurring of moving objects at the rain inpainting step.

After the removal of edge pixels, non-rain pixels are separated from the remaining candidate rain pixels. For this separation, the nature of the variations of intensity values of pixels in the consecutive frames are examined. Since intensity evolution of rain and non-rain pixels differ by the symmetry of the waveform, the following two features are selected, where $I_\ell(m, n)$ is the pixel at position (m, n) in l^{th} frame and $l \in 1, 2, \cdots, L$, where L is the block length of size $(2N + 1)$ and N is a positive integer.

1. *Range:* Difference between the maximum and minimum of the pixel intensity waveform is given by

$$range(m, n) = \max_\ell \left(I_\ell(m, n) \right) - \min_\ell \left(I_\ell(m, n) \right) \tag{5.6}$$

In rain regions, this difference is bound by a small threshold (say T_1), whereas, the feature *range* is based on the observation that the intensity variations caused by object motion is more than the changes caused by raindrop.

2. *Spread asymmetry:* Absolute difference of the standard deviation of pixel intensity waveform above the mean and below the mean in the rain region is bound by a small threshold (say T_2), where

$$spread\ asymmetry(m, n) = |A - B| \tag{5.7}$$

$$\text{where} \quad A = \operatorname*{std\ dev}_\ell \left(I_\ell(m, n) \mid I_\ell(m, n) > c \right)$$
$$B = \operatorname*{std\ dev}_\ell \left(I_\ell(m, n) \mid I_\ell(m, n) < c \right)$$
$$c = \operatorname*{mean}_\ell \left(I_\ell(m, n) \right)$$

Here feature *spread asymmetry* is exploiting the observation that the intensity variations in consecutive frames are more symmetric for rain pixels in comparison with non-rain pixels.

The above features are calculated within a block of $L = (2N + 1)$ consecutive frames, where N previous and N subsequent frames are considered. Distribution of pixels according to features *range* and *spread asymmetry* for rain and non-rain region for 15 consecutive frames of video with a static background ("pool") and dynamic background ("magnolia") are shown in Fig. 5.4 and 5.5 respectively. Results show that these two features are enough to discriminate rain pixels and non-rain moving object pixels.

For the discrimination of rain and non-rain pixels any classifier can be tried. Naive Bayes classifier [Duda et al., 2001] is a simple probabilistic classifier based on Bayes' theorem [Duda et al., 2001] with the assumption of independence of features. Naive Bayes classifier combines naive Bayes probability model with a decision rule as given below

$$v_{NB} = \operatorname*{arg\ max}_{\omega_j \in C} \ P(\omega_j) \prod_i P(\alpha_i / \omega_j) \tag{5.8}$$

Here α_i is the ith attribute and ω_j is the jth class. $P(\omega_j)$ is the prior probability and $P(\alpha_i / \omega_j)$ is the posteriori probability. $P(\alpha_i / \omega_j)$ is the probability that attribute α_i occurs for a pixel given the pixel belongs to class ω_j and C is the set of target. Here *range* and *spread asymmetry* are two

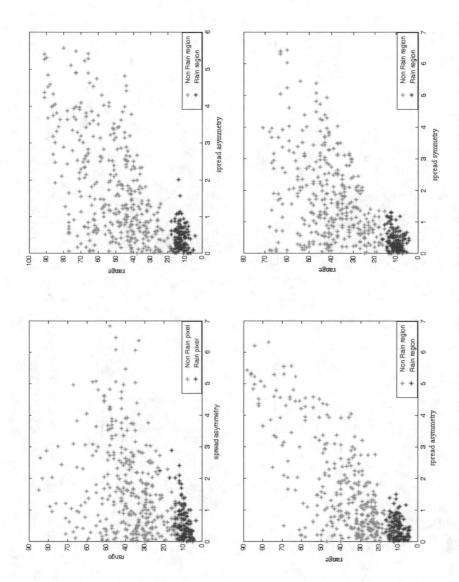

Figure 5.4: Distribution of pixels in feature space (*range* and *spread asymmetry*) for rain and non-rain region in 3D window ($w_x \times w_y \times w_t$) having dimension (a) $1 \times 1 \times 15$, (b) $5 \times 5 \times 15$, (c) $3 \times 3 \times 5$ and (d) $5 \times 5 \times 3$ for video with static background ("pool" video).

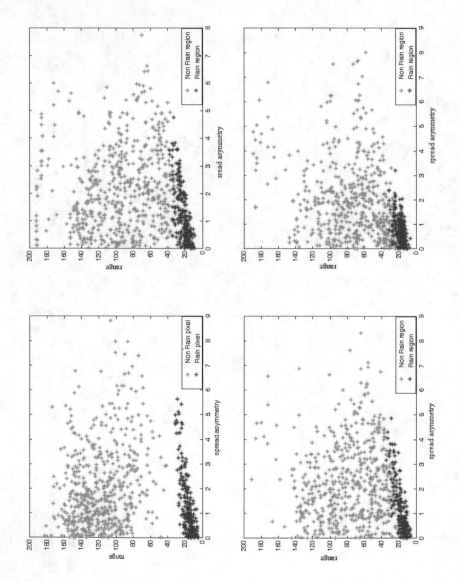

Figure 5.5: Distribution of pixels in feature space (*range* and *spread asymmetry*) for rain and non-rain region in 3D window ($w_x \times w_y \times w_t$) having dimension (a) $1 \times 1 \times 15$, (b) $5 \times 5 \times 15$, (c) $3 \times 3 \times 5$ and (d) $5 \times 5 \times 3$ for video with dynamic background ("magnolia" video).

attributes and *rain* and *non-rain* are two classes. Decision about the class (rain and non-rain) can be made according to the decision rule in eqn. (5.8).

Sum of the prior probability for the rain and non-rain pixels is unity. If there is a heavy rain then, the prior probability of the rain pixels is more than the prior probability of the non-rain pixels. If there is a light rain then, the prior probability of the rain pixels is less than the prior probability of the non-rain pixels. If there is no information then, rain and non-rain pixels can be considered equiprobable. For ease of implementation, the rain and non-rain pixels are considered equiprobable in this chapter. The *range* and *spread asymmetry* features are estimated from the same intensity waveform, and it is observed that these features are correlated. Here, for simplicity, these features are assumed to be independent of each other. Posterior probability distribution obtained from a particular frame is considered the same for all other video frames. Posterior probability of each attribute (feature) for rain and non-rain classes are shown in Fig. 5.6(a), 5.7(a), 5.8(a) and 5.9(a).

For simplicity two thresholds (T_1 and T_2) are selected based on the posterior probabilities for the discrimination of rain pixels from non-rain ones.

5.3.2   INPAINTING OF RAIN PIXELS

Intensity variations produced by raindrops are somewhat symmetric about the mean of the intensities of consecutive frames at a particular pixel position. Hence inpainting of detected rain pixels can be achieved by replacing it with the corresponding temporal mean of the intensity waveform.

5.4   SPATIOTEMPORAL DETECTION

Probabilistic temporal detection requires 15 consecutive (seven previous and seven next) frames for good accuracy. Examination of 15 consecutive frames increases the buffer size because buffer size increases proportionally with the number of frames. Large buffer size causes delay and adds to the cost of the system. Hence to reduce the buffer size and delay, number of frames required for the detection process should be less. But reducing the number of frames simultaneously reduces the number of pixels available for statistical inference, which affects the estimation accuracy. Hence for obtaining sufficient pixels for estimation, spatiotemporal detection is proposed in place of temporal detection process. Here the window under observation is extended in the spatial domain and thus the number of pixels under observation increases boosting the accuracy of statistical inference. Along with the increase in spatial window its span in the time domain is reduced, which means the detection process requires less number of frames. Thus, the spatiotemporal window provides sufficient pixels for accurate statistical estimate without increasing the requirement of the number of frames. In other words, for the spatiotemporal window, less number of frames (means less buffer size and delay) are required for the same detection accuracy.

Distribution of rain and non-rain pixels in feature space (*range* and *spread asymmetry*) for different spatiotemporal window size (3D window: ($w_x \times w_y \times w_t$)) are shown in Fig. 5.4 and 5.5, where w_x, w_y and w_t are window dimensions in x, y and *time* respectively. Prior and pos-

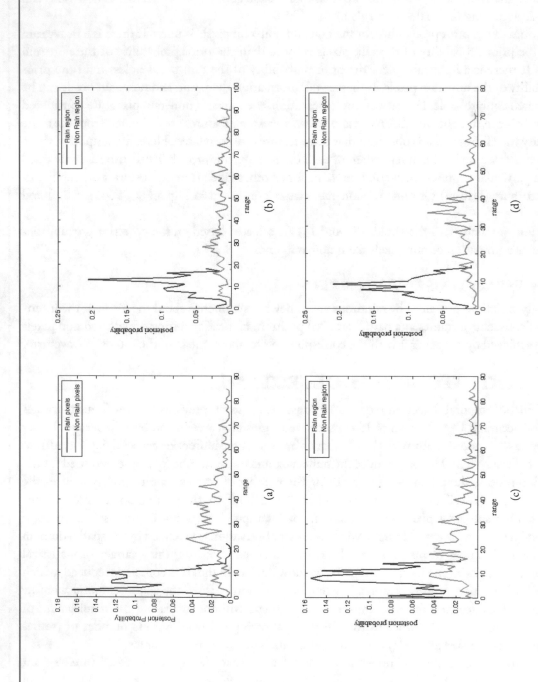

Figure 5.6: Posteriori probability of attribute *range* for rain and non-rain region in 3D window ($w_x \times w_y \times w_t$) having dimension (a) $1 \times 1 \times 15$, (b) $5 \times 5 \times 15$, (c) $3 \times 3 \times 5$ and (d) $5 \times 5 \times 3$ for video with static background ('pool' video).

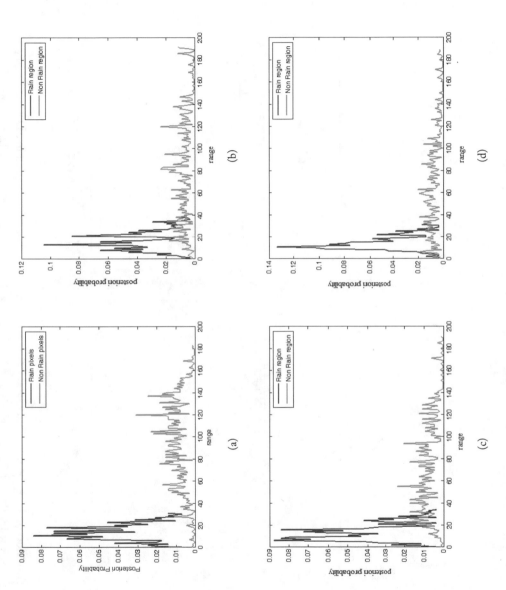

Figure 5.7: Posterior probability of attribute *range* for rain and non-rain region in 3D window ($w_x \times w_y \times w_t$) having dimension (a) $1 \times 1 \times 15$, (b) $5 \times 5 \times 15$, (c) $3 \times 3 \times 5$ and (d) $5 \times 5 \times 3$ for video with dynamic background ("magnolia" video).

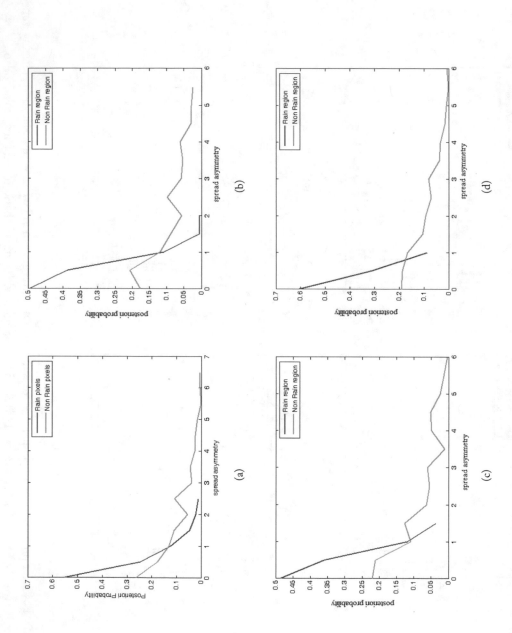

Figure 5.8: Posteriori probability of attribute *spread asymmetry* for rain and non-rain region in 3D window ($w_x \times w_y \times w_t$) having dimension (a) $1 \times 1 \times 15$, (b) $5 \times 5 \times 15$, (c) $3 \times 3 \times 5$ and (d) $5 \times 5 \times 3$ for video with static background ("pool' video).

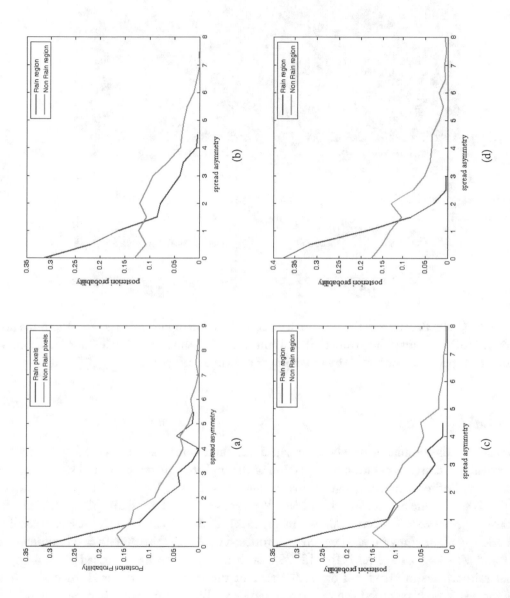

Figure 5.9: Posteriori probability of attribute *spread asymmetry* for rain and non-rain region in 3D window ($w_x \times w_y \times w_t$) having dimension (a) $1 \times 1 \times 15$, (b) $5 \times 5 \times 15$, (c) $3 \times 3 \times 5$ and (d) $5 \times 5 \times 3$ for video with dynamic background ("magnolia" video).

terior probabilities are estimated as described in temporal properties-based algorithm. Posterior probability of each attribute for rain and non-rain pixels are shown in Fig. 5.6, 5.7, 5.8 and 5.9.

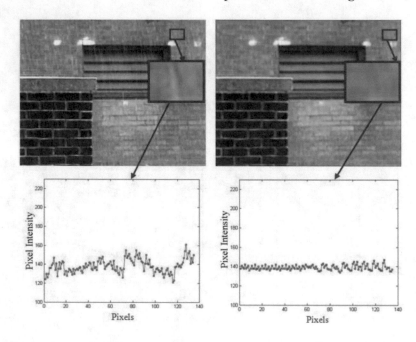

Figure 5.10: (Left) Frame from the "test" rain video and spatiotemporal intensity waveform at a pixel position; (Right) The same frame from the same rain video after rain removal by temporal rain removal technique and spatiotemporal intensity waveform at the same pixel position.

5.5   DATABASE

Videos used for the simulation are shown in Fig. 5.11. These videos have been used in the reported rain removal literature. Miss and false detection of the probabilistic temporal and spatiotemporal algorithms and other competing algorithms are calculated for the "black car01" [K. Garg and S. K. Nayar, 2006], "blue car" [S. Starik and M. Werman, 2003] and "football" [A. K. Tripathi and S. Mukhopadhyay, 2010] video as shown in Fig. 5.11(a), 5.11(b) and 5.11(c) respectively. Two rain-rendering algorithms are used which are proposed by A. K. Tripathi and S. Mukhopadhyay [2010] and K. Garg and S. K. Nayar [2006]. Both methods use image-based rendering, i.e., artificial rain streaks are rendered on each frame of the video. Tripathi et al. used the streak model and Garg & Nayar used the rain streak database. Binary difference between the original video frame and the rain-rendered video frame is taken as the ground truth for the calculation of the miss and false detection. In "black car01" video, rain is rendered by the method proposed in [K. Garg and S. K. Nayar, 2006]. In this video, a black car is moving, and heavy rain is rendered

Figure 5.11: Videos used for the simulation experiment with rain removal algorithms (a) "black car01" [K. Garg and S. K. Nayar, 2006], (b) "blue car" [S. Starik and M. Werman, 2003], (c) "football" [A. K. Tripathi and S. Mukhopadhyay, 2010], (d) "pool" [K. Garg and S. K. Nayar, 2004], (e) "magnolia" [K. Garg and S. K. Nayar, 2004], (f) "street" [K. Garg and S. K. Nayar, 2007] and (g) "street01" [K. Garg and S. K. Nayar, 2007].

over this video. This is a dynamic background outdoor video (frame size 376×504) captured by static camera. In "blue car" video, rain streaks are rendered randomly over a frame. Different rain video frames are generated by repeating this process over the frames. The frame size of this static background outdoor video "pool" is 240×320. In "football" video, rain is rendered by the method proposed in [A. K. Tripathi and S. Mukhopadhyay, 2010]. In this video (frame size 240×320), a football match is shown, and heavy rain is rendered over this video. This is a dynamic background outdoor video. Simulation is also performed over videos "pool" [K. Garg and S. K. Nayar, 2004], "magnolia" [K. Garg and S. K. Nayar, 2004], "street" [K. Garg and S. K. Nayar, 2007], and "street01" [K. Garg and S. K. Nayar, 2007] as shown in Fig. 5.11. In "pool" video rain streaks are shown with a wall in the background. This is a static background outdoor video and size of each frame is 275×196. In "magnolia" video (a clip from the movie "magnolia"), scene consists of a person moving and speaking on the phone and rain is visible through the glass window. The frame size of this dynamic background indoor video is 280×352. In "street" video (frame size 240×360), a traffic scene is shown where a yellow car is moving with a high speed. This is a dynamic background outdoor video with a moving object of high speed. In "street01" video, persons are moving in the street and raindrops can be seen over a large depth. This is a dynamic background outdoor video with a lot of moving objects and size of each frame is 240×360. In

Section 5.3.1, Fig. 5.4, 5.6 and 5.8 are derived from the "pool" video and Fig. 5.5, 5.7 and 5.9 derived from the "magnolia" video.

5.6 SIMULATION AND RESULTS

Simulation is performed in a MATLAB 7.0.4 environment. Probabilistic rain removal algorithms work only on the intensity plane. Hence prior to the detection and inpainting of rain pixels, input *RGB* frame is converted into YC_bC_r color space. Chrominance components (C_b and C_r) remain unchanged as shown in Fig. 5.2.

Results of the intermediate steps of the probabilistic algorithm are shown in Fig. 5.12. Results demonstrate the efficacy of the probabilistic temporal rain removal algorithm. In Fig. 5.12(c), removal of moving edge pixel suppresses a lot of false rain candidates. Fig. 5.12(d) shows that two attributes (*range* and *spread asymmetry*) are sufficient for the discrimination of rain from non-rain pixels.

Figure 5.12: Intermediate steps of the probabilistic temporal rain removal algorithm. (a) input frame, (b) all possible rain candidates (initial mask), (c) possible rain pixels after removal of edge pixels, (d) final rain mask (refinement by the attributes), (e) output rain removed frame.

Quantitative performance of the detection process is analyzed in terms of miss and false detection. Miss and false detection are averaged for 25 frames. Miss and false detection of the probabilistic rain removal algorithms are shown in Tables 5.2, 5.3 and 5.4. As the accuracy of the spatiotemporal rain detection algorithm is a function of window size, miss and false detection are calculated for various 3D window sizes. The size of 3D window also determines the number

Table 5.2: Average miss and false detection of probabilistic temporal and spatiotemporal rain detection algorithm for "black car01" video

Probabilistic Temporal			Window size in time domain							
Spatiotemporal			3	5	7	9	11	13	15
window size in spatial domain	1	(MD, FD)	(1222, 5181)	(762, 3465)	(317, 2401)	(260, 2397)	(253, 2173)	(244, 2124)	**(207, 2062)**
		Error	6403	4227	2718	2657	2426	2368	**2269**
	3	(MD, FD)	(1053, 3982)	(707, 3229)	(271, 2342)	(253, 2231)	(244, 2146)	(239, 2079)	**(216, 2051)**
		Error)	5035	3936	2613	2484	2390	2318	**2267**
	5	(MD, FD)	(980, 3931)	(727, 3153)	(272, 2338)	(265, 2275)	(279, 2177)	(278, 2123)	(247, 2108)
		Error	4911	3880	2610	2540	2456	2401	2325
	7	(MD, FD)	(974, 4010)	(752, 3161)	(293, 2359)	(292, 2341)	(291, 2313)	(287, 2222)	(255, 2201)
		Error	4984	3913	2652	2633	2604	2509	2456

Table 5.3: Average miss and false detection of probabilistic temporal and spatiotemporal rain detection algorithm for "blue car" video

Probabilistic Temporal Spatiotemporal			Window size in time domain						
			3	5	7	9	11	13	15
window size in spatial domain	1	(MD, FD)	(627, 4295)	(315, 2537)	(127, 1919)	(87, 1897)	(63, 1878)	(57, 1867)	**(47, 1822)**
		Error	4922	2952	2046	1984	1941	1924	**1869**
	3	(MD, FD)	(458, 3159)	(296, 2410)	(89, 1861)	(62, 1824)	(63, 1822)	(61, 1820)	**(54, 1819)**
		Error	3617	2706	1950	1886	1885	1881	**1873**
	5	(MD, FD)	(385, 3108)	(316, 2321)	(89, 1858)	(89, 1857)	(88, 1855)	(87, 1856)	(87, 1861)
		Error	3493	2637	1947	1946	1943	1943	1948
	7	(MD, FD)	(352, 3178)	(328, 2331)	(103, 1978)	(101, 1969)	(98, 1964)	(98, 1964)	(97, 1963)
		Error	3530	2659	2081	2070	2062	2062	2060

Table 5.4: Average miss and false detection of probabilistic temporal and spatiotemporal rain detection algorithm for "football" video

Probabilistic Temporal Spatiotemporal		Window size in time domain						
		3	5	7	9	11	13	15
window size in spatial domain	1	(MD, FD)	(1504,5668)	(1082,4357)	(941,3387)	(807, 3217)	(727, 2911)	(619, 2756)	**(522, 2698)**
		Error	7172	5439	4328	4024	3638	3375	**3220**
	3	(MD, FD)	(1663,4227)	(1262,3261)	(1127,3159)	(1009, 3159)	(825, 2892)	(755, 2792)	(665, 2768)
		Error)	5890	4523	4286	4168	3717	3547	3433
	5	(MD, FD)	(1758,4416)	(1348,3465)	(1212,3294)	(1079, 3215)	(978, 2987)	(873, 2913)	(809, 2880)
		Error	6174	4813	4506	4294	3965	3786	3689
	7	(MD, FD)	(1797,4580)	(1362,3468)	(1221,3355)	(1138, 3206)	(1015, 3090)	(912, 3024)	(894, 3003)
		Error	6377	4830	4576	4344	4105	3936	3897

Table 5.5: Average miss and false detection of competing algorithms for "black car01" video

Algorithms	MD	FD	Error	Error (p.u.)
Zhang et al.	631	2925	3556	1.57
Garg and Nayar	789	2683	3472	1.53
Liu et al.	697	2879	3576	1.58
Probabilistic temporal method ($w_t = 13$)	244	2124	2368	1.04
($w_t = 15$)	207	2062	2269	1
Spatiotemporal method ($5 \times 5 \times 3$)	980	3931	4911	2.16
($5 \times 5 \times 5$)	727	3153	3880	1.71
($5 \times 5 \times 7$)	272	2338	2610	1.15
($3 \times 3 \times 9$)	253	2231	2484	1.09
($3 \times 3 \times 11$)	244	2146	2390	1.05

Table 5.6: Average miss and false detection of competing algorithms for "blue car" video.

Algorithms	MD	FD	Error	Error (p.u.)
Zhang et al.	247	2236	2483	1.33
Garg and Nayar	274	1987	2261	1.21
Liu et al.	225	1962	2187	1.17
Probabilistic temporal method ($w_t = 13$)	57	1867	1924	1.03
($w_t = 15$)	47	1822	1869	1
Spatiotemporal method ($5 \times 5 \times 3$)	385	3108	3493	1.87
($5 \times 5 \times 5$)	316	2321	2637	1.41
($5 \times 5 \times 7$)	89	1858	1947	1.04
($3 \times 3 \times 9$)	62	1824	1886	1.01
($3 \times 3 \times 11$)	63	1822	1885	1.008

of frames required in the rain detection, which in turn determines the size of the buffer. Results shown in Tables 5.2, 5.3 and 5.4 have different values of figure of merit but follow the same trend. As the window size is increased in the time domain, miss and false detection (and Error) decrease but simultaneously the buffer size (proportional to the number of consecutive frames) and delay (frame acquisition time) increase. For a smaller temporal window, as the spatial window size is increased keeping the time domain window size constant, miss and false detection decrease from spatial dimension 1 to 3 but increase for 3 to 5 and more. Large temporal window (say 50 frames) is good only for static rain videos (i.e., no moving objects) captured by static camera. This is a naïve condition. If there are moving objects then a large time window will create more artifacts.

Table 5.7: Average miss and false detection of competing algorithms for "football" video

Algorithms	MD	FD	Error	Error (p.u.)
Zhang et al.	1004	4287	5291	1.64
Garg and Nayar	1169	3233	4402	1.37
Liu et al.	1117	3851	4968	1.54
Probabilistic temporal method ($w_t = 13$)	619	2756	3375	1.05
($w_t = 15$)	522	2698	3220	1
Spatiotemporal method ($3 \times 3 \times 3$)	1663	4227	5890	1.83
($3 \times 3 \times 5$)	1262	3261	4523	1.40
($3 \times 3 \times 7$)	1127	3159	4286	1.33
($3 \times 3 \times 9$)	1009	3159	4168	1.29
($3 \times 3 \times 11$)	825	2892	3717	1.15

A large time window is also not desirable for the real-time application due to large buffer size, cost and delay.

Results of miss and false detection for other competitive algorithms are shown in Tables 5.5, 5.6 and 5.7. The performance of rain detection degrades due to the presence of the dynamic objects irrespective of the detection technique used. Results show that the number of miss and false detection (and Error) for spatiotemporal algorithms are very low in comparison with most of the competing rain removal algorithms (Zhang et al., Garg and Nayar, and Liu et al.) and very close to the probabilistic temporal rain removal method. The reason for the a large error in the Garg and Nayar method is that heavy rain may corrupt two or more consecutive frames, and this method assumed that rain corrupts only one frame and very rarely two consecutive frames. One more reason is that variations in the size and velocity of the raindrops violate the photometric constraint, and the slow moving object also follows the photometric constraint of rain streaks. Thus, heavy rain increases miss detection and slow moving objects increase false detection. The method proposed by Zhang et al. gives a large error because slow moving objects also follow the chromatic property which increases false detection as noted in Tables 5.5, 5.6 and 5.7.

Liu et al. modified the photometric constraint. But still their assumptions are not valid in general. It increase the error in the detection process. Liu et al. requires three consecutive frames or two arbitrary frames other than the current frame for the detection of rain. The main drawback of the Liu et al. method is that if rain streaks cover the same location in two arbitrary frames then this method fails to detect rain streaks. This usually happens when there is a heavy rain because corruption due to heavy rain may persist for two or three consecutive frames. Thus in case of heavy rain, observation of five consecutive frames for the pixel-based rain detection method is more desirable. In the Liu et al. method, segmentation of rain and non-rain region is not optimum because several cases of false detection are observed at the boundaries separating

Table 5.8: Variance of a rain affected region in 15 consecutive frames of "pool" video

Win. Size	Method	Var.	Var. (p.u.)
1 × 1 × 15	Original	3.61	3.22
	Garg and Nayar	1.56	1.39
	Zhang et al.	1.55	1.38
	Liu et al.	1.59	1.40
	Probabilistic temporal method	1.12	1
	Spatiotemporal method	1.13	1.01
5 × 5 × 15	Original	13.92	1.52
	Garg and Nayar	11.89	1.30
	Zhang et al.	10.49	1.15
	Liu et al.	12.15	1.33
	Probabilistic temporal method	9.16	1
	Spatiotemporal method	9.16	1
11 × 11 × 15	Original	116.24	1.41
	Garg and Nayar	110.84	1.34
	Zhang et al.	89.18	1.08
	Liu et al.	109.55	1.33
	Probabilistic temporal method	82.62	1
	Spatiotemporal method	82.68	1.0007

the rain and non-rain regions. Thus, it can be concluded that the Liu et al. method fails when there is a heavy rain and/or various moving objects in the rain.

The probabilistic temporal method and spatiotemporal method give low miss detection in comparison with other competing algorithms. These probabilist algorithms have assumed that heavy rain may corrupt two or three consecutive frames which help in reducing the miss detection, and the use of class probabilities in classification helps in reducing the false detection. Furthermore, removal of edge pixels provides better aesthetic quality of frames.

The spatiotemporal approach enjoys the benefit of use of class probabilities in classification and edge pixels removal module in a similar way as the temporal approach. It can be concluded that for the performance of the rain removal process using spatiotemporal properties increases with the increase in the number of frames. For example video "Black car01" (5 × 5 × 3), (5 × 5 × 5) and (5 × 5 × 7) spatiotemporal windows provide 4991, 3880 and 2610 error pixels respectively along with savings in the buffer size and delay by 80%, 66.67% and 33.33% respectively. Though the price of quality degradations seems to be great along with the savings in the buffer size, it may be noted that the number of error pixels are only 2.59%, 2.04% and 1.38% of the full frame. Hence, a large saving in buffer size and delay can be obtained with little sacrifice in image quality. The use

Table 5.9: Variance of a rain-affected region in 15 consecutive frames of "magnolia" video

Window	Method	Var.	Var. (p.u.)
$1 \times 1 \times 15$	Original	24.09	7.95
	Garg and Nayar	4.41	1.46
	Zhang et al.	8.95	2.95
	Liu et al.	5.88	1.94
	Probabilistic temporal method	3.03	1
	Spatiotemporal method	3.05	1.01
$5 \times 5 \times 15$	Original	19.76	2.29
	Garg and Nayar	8.91	1.03
	Zhang et al.	11.92	1.38
	Liu et al.	12.25	1.42
	Probabilistic temporal method	8.62	1
	Spatiotemporal method	8.67	1.01
$11 \times 11 \times 15$	Original	114.22	8.63
	Garg and Nayar	14.83	1.12
	Zhang et al.	24.02	1.82
	Liu et al.	23.92	1.81
	Probabilistic temporal method	13.23	1
	Spatiotemporal method	13.29	1.01

of the spatiotemporal window though increases the number of samples in intensity waveform the increase in computational load is not much due to the use of simple features. Hence, the use of the spatiotemporal window provides flexibility to the designer to choose the right set of parameters for the system.

Simulations of probabilistic rain removal algorithms are carried out in videos with a static background ("pool" video) and dynamic or moving background ("magnolia," "street" and "street01") as shown in Fig. 5.11. Removal of rain from a video produces a smooth background. It means the rain removal process removes the intensity variations in consecutive frames. Sum of the prior probability for the rain and non-rain pixels is unity. If there is heavy rain, then prior probability of the rain pixels is more than the prior probability of the non-rain pixels. If there is a light rain, then prior probability of the rain pixels is less than the prior probability of the non-rain pixels. If there is no information then rain and non-rain pixels can be considered equiprobable. Posterior probability distribution obtained from a particular frame is considered the same for all other video frames. Quantitative performance of the removal process is analyzed in terms of variance. Variance is calculated over a rain-affected region with no moving objects in 15 consecutive frames within a 3D window of spatial size 1×1, 5×5 and 11×11. The lower the value of the variance the better the algorithm. Results for the variance are shown in Table 5.8 and Table 5.9 for the "pool" and "magnoliar" video respectively. Results show that the probabilistic spatiotemporal algorithm (3D window: $3 \times 3 \times 9$) gives low value of variance in comparison with other competing algorithms (Zhang et al. [Zhang et al., 2006], Garg and Nayar [K. Garg and S. K. Nayar, 2007, 2004] and Liu et al. [Liu et al., 2009]) and nearly similar to the probabilistic temporal method.

Qualitative results are shown in Fig. 5.13, Fig. 5.14, Fig. 5.15 and Fig. 5.16. Results show that the probabilistic temporal algorithm removes rain more effectively than other competing rain removal algorithms in terms of the perceptual image quality. Other rain removal algorithms produce some artifacts. In the "magnolia" video, the degradation is more visible near the fingers of a man (see Fig. 5.14). In "street" video, degradation can be seen over the yellow car (see Fig 5.15). In "street01" video, degradations are visible over the hand shown in the right part of the video (see Fig. 5.16). The probabilistic spatiotemporal algorithm produces the same visual quality as produced by the probabilistic temporal method (i.e., no degradations in image quality) but with the advantage of less buffer size and delay. Video results can be found at http://www.ecdept.iitkgp.ernet.in/web/faculty/smukho/docs/spatiotemporal/rain_spatiotemporal.html.

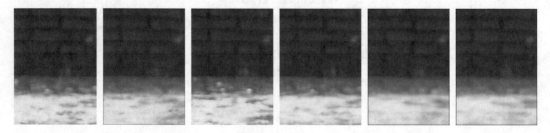

Figure 5.13: (a) Original "pool" rain video (frame 40). The same frame after rain removal by (b) the Zhang et al. method, (c) the Garg and Nayar method, (d) the Liu et al. method, (e) the probabilistic temporal method and (f) the spatiotemporal method.

5.7   CONCLUSION

In this chapter, novel, efficient and probabilistic model-based temporal and spatiotemporal rain removal algorithms are presented. It was found that rain and non-rain regions have different time evolution properties of the pixel intensity. Probabilistic algorithms use this property to separate rain pixels from non-rain pixels. In the probabilistic temporal rain removal algorithm, the discrimination process requires a large number of consecutive frames that causes large buffer size and delay. Hence, a spatiotemporal approach is presented. This new spatiotemporal approach reduces the buffer size requirement and delay. Thus, the probabilistic spatiotemporal algorithm paves the way for the real-time implementation. Quantitative and qualitative results show that the probabilistic algorithms remove rain effectively in comparison with most of the competing

Figure 5.14: (a) Original "magnolia" rain video (frame 45). The same frame after rain removal by (b) the Zhang et al. method, (c) the Garg and Nayar method, (d) the Liu et al. method, (e) the probabilistic temporal method and (f) the spatiotemporal method.

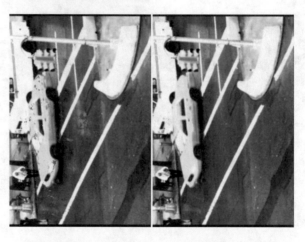

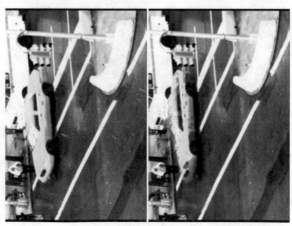

Figure 5.15: (a) Original "street" rain video (frame 81). The same frame after rain removal by (b) the Zhang et al. method, (c) the Garg and Nayar method, (d) the Liu et al. method, (e) the probabilistic temporal method and (f) the spatiotemporal method.

Figure 5.16: (a) Original "street01" rain video (frame 50). The same frame after rain removal by (b) the Zhang et al. method, (c) the Garg and Nayar method, (d) the Liu et al. method, (e) the probabilistic temporal method and (f) the spatiotemporal method.

rain removal algorithms. Probabilistic algorithms work only on the intensity plane. Thus the use of a single plane reduces the complexity of the algorithm. Probabilistic algorithms do not assume the shape, size and velocity of raindrops that make it robust to different rain conditions. In summary, the probabilistic temporal algorithm outperformed the competition in accuracy, and the probabilistic spatiotemporal algorithm outperformed the competition in overall quality.

CHAPTER 6

Impact of Camera Motion on Detection of Rain

6.1   INTRODUCTION

Rain removal algorithms require a certain number of consecutive frames for reducing the visibility of rain in videos. These algorithms are designed for the rain videos captured by a static camera. In this chapter, it is demonstrated that using global motion estimation all the rain removal algorithms developed for videos captured by static camera can be used for videos captured by a moving camera.

In Section 6.2 and Section 6.3 the probable rain candidates and the estimation of global motion parameters are explained. After global motion compensation, all rain removal algorithms developed for videos captured by a static camera can be used for videos captured by a moving camera. Few images captured by camera in motion are shown in Section 6.4. Simulation and results are shown in Section 6.5. Section 6.6 concludes this chapter.

6.2   RAIN CANDIDATES SELECTION

In the existing rain removal algorithms [K. Garg and S. K. Nayar, 2007, 2004, Zhang et al., 2006, Liu et al., 2009, N. Brewer and N. Liu, 2008] it is considered that due to rain, the same pixel may be corrupted in one or more consecutive frames. Schematic and pictorial views of the intensity changes of the five consecutive frames (current frame with two previous and two successive frames) are shown in Fig. 5.3.

This rain candidates selection process is valid only for the static video camera. When there is a relative motion between the camera and the objects, the location of the object pixels change in the subsequent frames. To track the accurate pixel position in previous and next frames, it is necessary to find out the global motion parameters of the frames under consideration with respect to the current frame.

6.3   GLOBAL MOTION PARAMETERS

Here a geometric model is used to characterize the global motion [A. M. Tekalp, 1995, T. Chen, 2000, Su et al., 2005]. We have assumed that there is no image warping. Hence, instead of a projective model a geometric model is chosen. In the case of rain video, the image or frame motion is attributed by the rotation and translation of the camera. Thus the following geometric model is applicable.

$$\begin{bmatrix} x' \\ y' \end{bmatrix} = \begin{bmatrix} \alpha & -\beta \\ \beta & \alpha \end{bmatrix} \begin{bmatrix} x \\ y \end{bmatrix} + \begin{bmatrix} x_{off} \\ y_{off} \end{bmatrix} \tag{6.1}$$

where (x, y) is the pixel position in the current frame and (x', y') is the pixel position in the target frame and $\phi = \arctan(\frac{\beta}{\alpha})$ represents the angle of rotation. Here, it is assumed that camera motion is limited to translation and rotation. The rotations may be due to the handshaking. Parameters α and β are used for this purpose. The motion due to translation are captured with the offset parameters (i.e., x_{off} and y_{off}). Since modern rain removal algorithms are using a limited number (viz., 5 to 15) of consecutive frames, zooming and scaling can be neglected. For the restoration of video, this geometric model is found to be simple and adequate.

To determine the global transformation parameters, first the video frame is divided into number of blocks. Using a motion estimation algorithm, the block motion vectors are estimated. The pixel corresponding to the block center is selected based on the local block motion vector. To obtain the global motion parameters α, β, x_{off} and y_{off}, the center of the blocks and their corresponding points in the reference frame are arranged in eqn. (6.2).

$$\begin{bmatrix} x'_1 \\ y'_1 \\ x'_2 \\ y'_2 \\ \vdots \\ x'_N \\ y'_N \end{bmatrix} = \begin{bmatrix} x_1 & -y_1 & 1 & 0 \\ y_1 & x_1 & 0 & 1 \\ x_2 & -y_2 & 1 & 0 \\ y_2 & x_2 & 0 & 1 \\ \vdots & & & \\ x_N & -y_N & 1 & 0 \\ y_N & x_N & 0 & 1 \end{bmatrix} \begin{bmatrix} \alpha \\ \beta \\ x_{off} \\ y_{off} \end{bmatrix} \tag{6.2}$$

Eqn. (6.2) describes an over-determined system. The total least square-singular value decomposition (TLS-SVD) is used for its solution [I. Markovsky and S. V. Huffel, 2007].

Assume $A\Phi \approx B$, where $A \in \mathfrak{R}^{m \times n}$ and $B \in \mathfrak{R}^{m \times d}$ are known and $\Phi \in \mathfrak{R}^{n \times d}$ is unknown.

Let $C = [A\ B] = U\Sigma V^T$ be the SVD of C, where $\Sigma = diag(\sigma_1, \ldots, \sigma_{n+d})$; $\sigma_1 \geq \ldots \geq \sigma_{n+d}$ be singular values of C. The singular vector space and singular values may be partitioned as

$$V = \begin{bmatrix} V_{11} & V_{12} \\ V_{21} & V_{22} \end{bmatrix} \quad \text{and} \quad \Sigma = \begin{bmatrix} \Sigma_1 & 0 \\ 0 & \Sigma_2 \end{bmatrix}$$

where $\Sigma_1 = diag(\sigma_1, \ldots, \sigma_n)$, $\Sigma_2 = diag(\sigma_{n+1}, \ldots, \sigma_{n+d})$. A TLS solution exists iff the matrix V_{22} (of size $(d \times d)$) is non-singular. The solution is unique iff $\sigma_n \neq \sigma_{n+1}$ and it is given by

$$\hat{\Phi}_{tls} = -V_{12}V_{22}^{-1} \tag{6.3}$$

where $\hat{\Phi}_{tls} = [\alpha\ \beta\ x_{off}\ y_{off}]^T$ are the global motion parameters in eqn. (6.3). If the pixel position in the current frame is (x, y) then the corresponding pixel position (x', y') in the target

(a)   (b)

Figure 6.1: Videos used for the simulation (a) "black car02" [K. Garg and S. K. Nayar, 2006], (b) "window" [K. Garg and S. K. Nayar, 2006].

frame can be calculated with the help of the global motion parameters. To find the pixel values at integer grid, bilinear interpolation is used. Once the motion-compensated frames are obtained, it is easy to find out the possible rain candidates or streaks by examining the intensity changes in consecutive frames. Due to the presence of the moving objects this rain selection process results in some false positives. Thus the output of the rain selection process requires some refinement.

6.4   DATABASE

A simulation experiment is carried out with the help of videos shown in Fig. 6.1. Most of the video sequences are reported earlier. In "black car02" video, a black car is moving in the heavy rain. This is a dynamic background outdoor video captured by a moving camera (this video clip is different from the "black car01" video discussed earlier). The "window" video is a static background outdoor video captured by a moving camera. In videos "black car02" and "window" rain is rendered by the method proposed in [K. Garg and S. K. Nayar, 2006].

6.5   SIMULATION AND RESULTS

Simulation is performed in a MATLAB 7.0.4 environment. All possible rain candidates can be properly detected if consecutive frames are aligned. If there is a movement of the camera with respect to the scene then difference of the consecutive frames produces the object edges which gives undesirable shape to the rain streaks and degrades the performance of the detection process. The advantage of the global motion compensation is shown in Fig. 6.2. Result shows that global

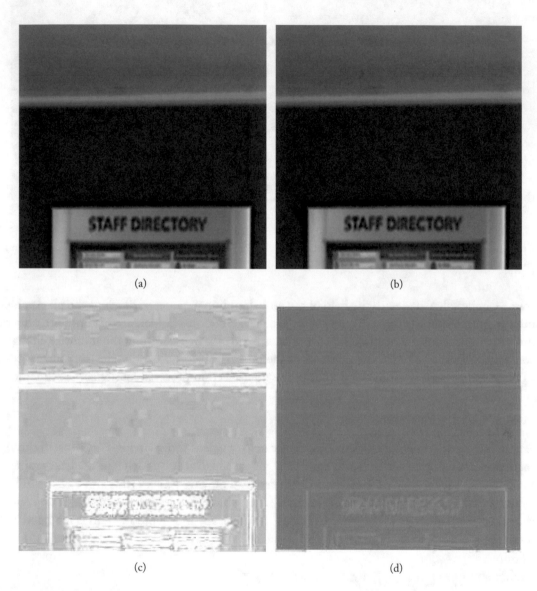

(a)

(b)

(c)

(d)

Figure 6.2: (a) and (b) Two consecutive frames taken by the moving camera; difference of these consecutive frames (c) before and (d) after global motion compensation. Here (c) and (d) are scaled appropriately for the display.

motion compensation accurately aligns the consecutive frames without producing any object edge for static objects in the difference image of the consecutive frames. Thus, estimation of the global motion parameters paves the way for the removal of rain when the video camera is in motion.

This global motion compensation (GMC) is expected to improve the performance of any rain removal algorithms. The Barnum et al. algorithm is one of the important algorithms which can take care of camera motion. It is found that the Barnum et al. algorithm performs better in comparison with other competing algorithms which are not designed for the camera in motion. The GMC applied prior to various states of the art algorithms, and it is expected that the rain removal accuracy of these algorithms will improve. The rain removal accuracy is calculated based on two percentage measures of the amount of the rain removed (H) and the amount of the scene distortion (E) [Barnum et al., 2007]. High value of H and low value of E indicate better performance. Results of competing rain removal algorithms are shown in Table 6.1. We have compared the performance of various state-of-the-art algorithms (Barnum et al., Garg & Nayar, Liu et al., probabilistic temporal and spatiotemporal) with and without GMC. The method proposed by Zhang et al. [2006] is not included in this comparison because it requires all the video frames for the removal of rain and GMC can not be successfully applied on all the frames. Low performance of the Zhang et al. method as described in section 5.6 is also a reason for leaving it out of competition.

Results in Table 6.1 show that the addition of global motion compensation improves the performance significantly (percentage of rain removal increases and scene distortion decreases) for all the algorithms. Results are evaluated over the 'black car02' and "window" videos [K. Garg and S. K. Nayar, 2006] (see Fig. 6.1). These said videos are captured by a moving camera. In these videos, rain is rendered by the method proposed in [K. Garg and S. K. Nayar, 2006]. Table 6.1 shows that without the additional step of GMC, performance of Garg & Nayar and Liu et al. methods are low. However, with the addition of GMC, performance of these methods (Garg & Nayar and Liu et al.) increases significantly and reaches close to the Barnum et al. method. In the case of temporal and spatiotemporal algorithms, even without GMC, their performances are far superior to Barnum et al. in terms of rain removal. Only the scene distortions are slightly high. With the addition of GMC, performance of temporal and spatiotemporal algorithms improves by a significant amount. The percentage of rain removed using probabilistic temporal and spatiotemporal technique are 41 to 69% higher than competition and scene distortions are very close to Barnum et al. Thus, it can be concluded that addition of GMC, improves the performance of all rain removal algorithms designed for a static camera and can be applied for videos captured by a camera in motion.

The Barnum et al. method is applicable for the videos captured by the moving and/or fixed camera. That's why the additional step of GMC is not included. The Barnum et al. method removes the rain effectively without any unwanted blurring. Here a blurred Gaussian model is used to approximate the rain streak. But the Barnum et al. method is suitable only for those rain streaks which are prominent. The Barnum et al. method fails to detect and remove the rain streak when

Table 6.1: Comparison for the percentage of rain removed (H) and scene distortions (E) caused by the state-of-the-art rain removal algorithms with and without GMC

Video	Method									
	Barnum et al.		Garg & Nayar (without GMC)		Liu et al. (without GMC)		Temporal (without GMC)		Spatiotemporal (without GMC)	
	$H(\%)$	$E(\%)$	$H(\%)$	$E(\%)$	$H(\%)$	$E(\%)$	$H(\%)$	$E(\%)$	$H(\%)$	$E(\%)$
"black car02"	41.11	6.59	36.17	7.12	37.31	7.03	52.89	8.11	52.91	7.01
"window"	31.01	0.36	27.13	0.55	27.87	0.49	47.71	0.74	48.56	0.48

Video	Method									
	Barnum et al.		Garg & Nayar (with GMC)		Liu et al. (with GMC)		Temporal (with GMC)		Spatiotemporal (with GMC)	
	$H(\%)$	$E(\%)$	$H(\%)$	$E(\%)$	$H(\%)$	$E(\%)$	$H(\%)$	$E(\%)$	$H(\%)$	$E(\%)$
"black car02"	41.11	6.59	40.29	6.87	40.43	6.74	58.08	7.51	58.46	6.71
"window"	31.01	0.36	30.18	0.44	30.24	0.37	51.11	0.63	52.45	0.44

it is not sharp enough because less prominent rain streaks do not create noticeable changes in the frequency domain. In the Barnum et al. method, rain streaks are considered globally consistent in orientation. Thus rain streaks located in other orientations (deviation in orientation caused by the wind) are missed.

The performance of the Garg & Nayar method with the addition of motion compensation reaches very close to the Barnum et al. method. Liu et al. modified the photometric constraint but still their assumption is not always valid leading to error in the detection process. The Liu et al. method requires three consecutive frames or two arbitrary frames for the detection of the rain. The main drawback of the Liu et al. method is that if rain streaks cover the same location in two arbitrary frames then this method fails to detect rain streaks. This usually happens when there is heavy rain because corruption due to heavy rain may last up to three consecutive frames. Thus in case of heavy rain, observation of five consecutive frames for the pixel-based rain detection method is more desirable. Due to the large number of failures in detection, it results in the low percentage of rain removal but low latency (three consecutive frames) results in low scene distortions.

Probabilistic temporal and spatiotemporal algorithms (with and without GMC) have high percentage of rain removal. One of the reasons for this high performance is the assumption of the rain corruption that may last up to five consecutive frames. This assumption helps in the detection of all possible rain candidates. Since the latency of the temporal and spatiotemporal algorithms are fifteen and nine consecutive frames respectively, these algorithms give low scene distortions in comparison to the Garg & Nayar method (having latency of thirty consecutive frames). It is clear from the results that all the state-of-the-art algorithms (which were applicable for the videos captured by the fixed camera) can be applied for the videos captured by camera in motion, if in the detection stage global motion compensation is added. However, this additional step increases the complexity of these algorithms. If rain videos are compressed using any video compression standard then we can easily get the information about the local motion from the encoder without any additional computation. The global motion parameters can be computed using these local motion parameters.

6.6   CONCLUSION

In this chapter the concept of global motion [T. Chen, 2000] is introduced in rain removal algorithms. It is shown that, using global motion compensation, all the rain removal algorithms developed for static camera can be used for videos captured by camera in motion to achieve better performance.

CHAPTER 7

Meteorological Approach for Detection and Removal of Rain from Videos

7.1 INTRODUCTION

In this chapter a novel, efficient and simple algorithm for detection and removal of rain from video using meteorological properties is presented. Here meteorological properties of rain are used to separate the rain pixels from the non-rain pixels. The said meteorological algorithm achieves good accuracy in spite of less number of consecutive frames that reduce the buffer size and delay. It works only on the intensity plane which further reduces the complexity and execution time significantly.

This chapter is organized as follows. In Section 7.2, the relation between the wind velocity and orientation of rain streak is developed. Here shape features of the rain streaks are also calculated. In Section 7.3, an algorithm to reduce the visibility of rain is presented. Summary of the algorithm is explained in Section 7.4. In Section 7.5 videos used for simulation are discussed. Simulation and results are shown in Section 7.6. Section 7.7 concludes this chapter.

7.2 STATISTICAL MODEL OF RAINDROP

A statistical model of raindrop provides deep understanding of the effect of rain. Here the effect of the wind velocity on the orientation of rain, shape and size of the rain streaks is investigated. The information on how a rain streak appears, combined with the estimation of the range of streak sizes, allow one to estimate the appearance of rain in the video.

7.2.1 DYNAMICS OF RAINDROP

Here the relation between the wind velocity and the direction of the rainfall is derived. It is assumed that wind and gravity are the two factors that influence the motion of the raindrops [Feng et al., 2006]. Consider a spherical raindrop of mass m and radius R falling under gravity g with a velocity v, while wind velocity is u. This raindrop is subjected to a buoyant force [G. K. Batchelor, 2000] $F_B = \rho_a V g$, where ρ_a is the density of air, V is the volume of the raindrop and

g is the acceleration due to gravity. A drag force [G. K. Batchelor, 2000] $F_d = 6\pi\mu Rv$, where μ is the dynamic viscosity, will also act opposite to the direction of propagation as shown in Fig. 7.1.

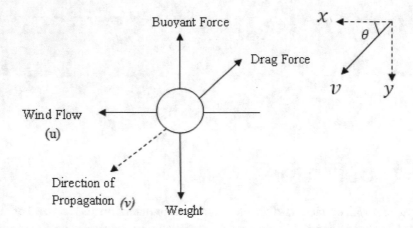

Figure 7.1: Free-body diagram of the falling raindrop.

If a_x and a_y are the acceleration, acting on the horizontal and vertical direction respectively, then

Horizontal forces

$$\sum F_x = ma_x$$

$$ma_x = 6\pi\mu R(u - v_x) \tag{7.1}$$

where v_x is the horizontal component of the raindrop velocity.

Vertical forces

$$\sum F_y = ma_y$$

$$ma_y = mg - \rho_a Vg - 6\pi\mu Rv_y \tag{7.2}$$

From eqn. (7.1)

$$a_x = \frac{6\pi\mu R}{m}(u - v_x)$$

or,

$$\frac{dv_x}{dt} = \frac{6\pi\mu R}{m}(u - v_x)$$

or,

$$\int_0^{v_x} \frac{dv_x}{\frac{6\pi\mu R}{m}(u - v_x)} = \int_0^t dt$$

or,

$$v_x = u[1 - e^{-\frac{6\pi\mu R}{m}t}] \tag{7.3}$$

From eqn. (7.2)

$$a_y = g - \frac{\rho_a Vg}{m} - \frac{6\pi\mu R}{m}v_y$$

or,

$$\frac{dv_y}{dt} = g - \frac{\rho_a Vg}{m} - \frac{6\pi\mu R}{m}v_y$$

or,

$$\int_0^{v_y} \frac{dv_y}{(g - \frac{\rho_a Vg}{m}) - \frac{6\pi\mu R}{m}v_y} = \int_0^t dt$$

or,

$$v_y = \frac{m}{6\pi\mu R}(g - \frac{\rho_a Vg}{m})[1 - e^{-\frac{6\pi\mu R}{m}t}] \tag{7.4}$$

Rainfall direction is

$$\theta = \tan^{-1}(\frac{v_y}{v_x}) \tag{7.5}$$

From eqn. (7.3) and eqn. (7.4) rainfall direction is

$$\theta = \tan^{-1}[\frac{m}{6\pi\mu R}(g - \frac{\rho_a Vg}{m})\frac{1}{u}] \tag{7.6}$$

Eqn. (7.6) shows how the orientation of the rainfall depends on the wind velocity. A free-falling object achieves its terminal velocity [R. Gunn and G. D. Kinzer, 1949] when the net force on the object is zero. Hence, resulting acceleration is zero for the object falling with terminal velocity. Thus the raindrop achieves its terminal velocity when $ma_x = 0$ and $ma_y = 0$. Hence from eqn. (7.1) and (7.2), terminal velocity is:

$$v_x = u \tag{7.7}$$

and

$$v_y = \frac{mg - \rho_a Vg}{6\pi\mu R} \tag{7.8}$$

If ν is the kinematic viscosity [G. K. Batchelor, 2000] then

$$\nu = \frac{\mu}{\rho_a} \tag{7.9}$$

Reynolds number [G. K. Batchelor, 2000] is defined as

$$R_e = \frac{2Rv_y}{\nu} \tag{7.10}$$

From eqn. (7.9) and (7.10)

$$\mu = \frac{2Rv_y\rho_a}{R_e} \tag{7.11}$$

Substitute the value of μ in eqn. (7.8)

$$v_y^2 = \frac{R_e(mg - \rho_a Vg)}{12\pi\rho_a R^2} \tag{7.12}$$

If ρ_s is the density of raindrop then

$$m = V\rho_s = \frac{4}{3}\pi R^3 \rho_s$$

Substituting the value of m and V in eqn. (7.12)

$$v_y = \sqrt{\frac{R_e g}{9}\frac{(\rho_s - \rho_a)}{\rho_a}}R \tag{7.13}$$

Thus from eqn. (7.5), eqn. (7.7) and eqn. (7.13) orientation is

$$\theta = \tan^{-1}\left[\frac{1}{u}\sqrt{\frac{R_e g}{9}\frac{(\rho_s - \rho_a)}{\rho_a}}R\right] \tag{7.14}$$

Terminal velocity

$$v_T = \sqrt{(v_x)^2 + (v_y)^2} \tag{7.15}$$

From eqn. (7.7) and eqn. (7.13)

$$v_T = \sqrt{u^2 + \frac{R_e g}{9}\frac{(\rho_s - \rho_a)}{\rho_a}R} \tag{7.16}$$

If there is no wind flow then $u = 0$ and thus terminal velocity v_T is

$$v_T = \sqrt{\frac{R_e g}{9}\frac{(\rho_s - \rho_a)}{\rho_a}}R \tag{7.17}$$

Based on the eqn. (7.17), terminal velocity at different values of the radius of raindrop in presence of wind and absence of wind is shown in Fig. 7.2. The resistance of air limits the speed of raindrop. Similar results (0.18–9.17 m/s) are reported by Gunn et al. [R. Gunn and G. D. Kinzer, 1949].

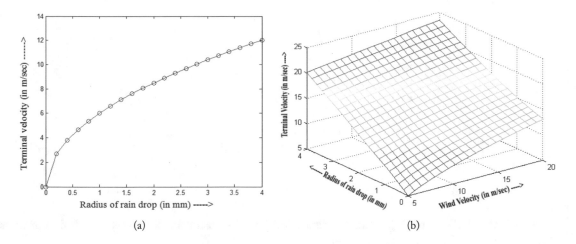

Figure 7.2: The plot of terminal velocity of raindrop against raindrop radius (a) in absence of wind and (b) in presence of wind.

7.2.2 3D TO 2D PROJECTION

The world is in 3D, and the images captured by the camera are in 2D. Hence 3D to 2D mapping is inevitable. Due to the high velocity of raindrops relative to the exposure time of the camera, it generates rain streaks in the image plane. Visual properties of rain streaks can be analyzed by the 3D to 2D projective camera model [R. C. Gonzalez and R. E. Woods, 1992]. If (x, y, z) is the camera coordinates, f is the focal length of the lens and (X, Y, Z) is the world coordinates then it can be represented by

$$\begin{bmatrix} x \\ y \\ z \end{bmatrix} = \begin{bmatrix} \frac{fX}{(f-Z)} \\ \frac{fY}{(f-Z)} \\ \frac{fZ}{(f-Z)} \end{bmatrix} \tag{7.18}$$

Since on the image plane z is zero by choice, consider only (x, y). Hence

$$x = \frac{fX}{(f - Z)} \quad \text{and} \quad y = \frac{fY}{(f - Z)} \tag{7.19}$$

If T is the exposure time of the camera then in 3D space a raindrop moves from coordinate (X, Y, Z) to coordinate $(X + v_x T, Y + v_y T, Z)$ within that interval. Hence in 2D space this movement recorded as $(\frac{fX}{f-Z}, \frac{fY}{f-Z})$ to $(\frac{f(X+v_x T)}{f-Z}, \frac{f(Y+v_y T)}{f-Z})$.

As shown in Fig. 7.3(a), the vertical component of rain streak size $d_y = \frac{f}{f-z}(2R + v_y T)$, the horizontal component of rain streak size $d_x = \frac{f}{f-z}(2R + v_x T)$ and "angle of incidence" $\alpha = \arctan(\frac{d_x}{d_y})$, where R is the radius of the raindrop. Thus the length and breath of the streak can

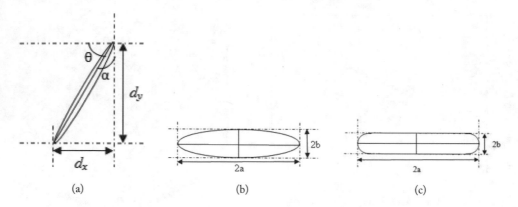

Figure 7.3: (a) Rain streak formation orientation and dimension in 2D projection. Rain streak as (b) an ellipse, and (c) a rounded rectangle.

be denoted as

$$\text{length} = \frac{f}{f-Z}(2R + v_y T)\sec\alpha \quad \text{and} \quad \text{breath} = \frac{f}{f-Z}2R \qquad (7.20)$$

Considering the variations of real world phenomena, rain streaks will have a variation in the exact shape. Thus two different types of rain streak shapes are assumed. Consider rain streak as an ellipse with major axis $2a$ and minor axis $2b$ as shown in Fig. 7.3(b). Area and aspect ratio of the rain streak can be denoted as follows.

$$\text{Area} = \pi ab = \frac{\pi}{4}\left(\frac{f}{f-Z}(2R + v_y T)\sec\alpha\right)\left(\frac{f}{f-Z}2R\right) \qquad (7.21)$$

$$\text{Aspect Ratio} = \frac{b}{a} = \frac{\frac{f}{f-Z}2R}{\frac{f}{f-Z}(2R + v_y T)\sec\alpha} = \frac{2R}{(2R + v_y T)\sec\alpha} \qquad (7.22)$$

Considering rain streak as a rounded rectangle with major axis $2a$ and minor axis $2b$ as shown in Fig. 7.3(c). Area and aspect ratio of the rain streak can be denoted as follows.

$$\text{Area} = 4ab + (\pi - 4)b^2 = \left(\frac{f}{f-Z}(2R + v_y T)\sec\alpha\right)\left(\frac{f}{f-Z}2R\right) + (\pi - 4)\left(\frac{f}{f-Z}R\right)^2 \qquad (7.23)$$

$$\text{Aspect Ratio} = \frac{b}{a} = \frac{\frac{f}{f-Z}2R}{\frac{f}{f-Z}(2R + v_y T)\sec\alpha} = \frac{2R}{(2R + v_y T)\sec\alpha} \qquad (7.24)$$

It is interesting to note that aspect ratio is independent of the shape of raindrop.

7.3   DETECTION AND REMOVAL OF RAIN STREAKS

Here meteorological properties are used for the identification of the rain streaks. First all possible rain candidates are identified by the intensity changes of N_l consecutive frames. Here it is assumed $N_l = 5$. Then rain candidates are finalized using meteorological property (shape and orientation). In this step, shape features (area and aspect ratio) and orientation of each possible rain candidates are examined to find out the potential rain candidates.

7.3.1   RAIN CANDIDATES SELECTION

In previous algorithms [K. Garg and S. K. Nayar, 2007, 2004, Zhang et al., 2006, Liu et al., 2009, N. Brewer and N. Liu, 2008] it is considered that due to rain, the same pixel may be corrupted in one or two consecutive frames. Here, to increase the performance of the detection process, similar to the probabilistic temporal and spatiotemporal methods, it is considered that heavy rain may affect up to three consecutive frames in a rain video. Schematic and pictorial views of the intensity changes of the five consecutive frames (current frame with two previous and two successive frames) are shown in Fig. 5.3.

This rain candidates selection process is valid only for the static video camera. In the case of relative motion between the camera and the objects, the pixel location of the object changes in the subsequent frames. To track the accurate pixel positions in subsequent frames, it is necessary to apply global motion compensation as explained in Chapter 6.

7.3.2   IDENTIFYING POTENTIAL RAIN CANDIDATES

According to the statistical model as discussed in Section 7.2, each rain streak has a particular area, aspect ratio and orientation.

According to the meteorological data set,

$$\text{average wind velocity } u = 3.4\,kmph = v_x$$
$$\approx 1m/sec$$

$$\text{and raindrop velocity } v_y \approx 2 - 10m/sec$$

where $\rho_s = 998.2071\,kg/m^3$, $\rho = 1.2041\,kg/m^3$, $g = 9.8\,m/sec^2$ and $R_e = 40$ for water

Assuming exposure time $T = 1\,ms - 40\,ms$, and radius of raindrop $R = 0.1\,mm - 3\,mm$ [K. Garg and S. K. Nayar, 2007] and substituting the values of all the variables in the eqn. (7.14), (7.22) and (7.24), we get

$$\text{Orientation } \quad \theta = (5° - 30°)$$
$$\text{Aspect Ratio} \approx \frac{1}{4} \text{ to } \frac{1}{96} \approx 0.25 - 0.012$$

The estimated range conforms to the reference [N. Brewer and N. Liu, 2008]. Area of rain streak is empirically found to be in between 10 and 800 pixels. Among all possible rain candidates, there are lots of false positives due to the motion of objects. For the refinement of rain candidates area, aspect ratio and orientation of the possible rain candidates are computed. Those candidates, for which the values fall within the admissible range, are marked as the true rain candidates.

7.3.3   INPAINTING OF RAIN PIXELS

Inpainting of each rain pixels in the nth frame with intensity I_n can be achieved by the temporal mean of the I_{n-2} to I_{n+2} pixel intensities. If there is a relative motion between the video camera and the scene, then it is necessary to compensate the motion using the global motion parameters before inpainting.

7.4   SUMMARY OF THE ALGORITHM

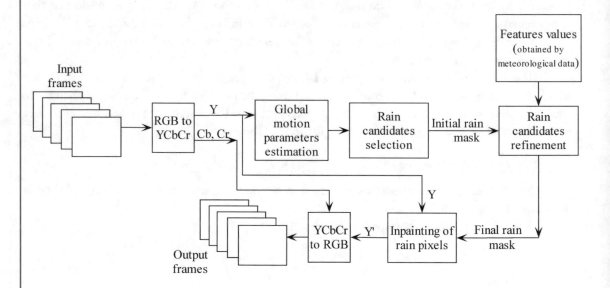

Figure 7.4: Block diagram of the meteorological rain detection and removal algorithm for video captured by moving camera.

In order to reduce the effect of rain in the videos, it is necessary to find out the rain pixels and then replace these pixels with the estimated non-rain pixel values. Block diagram of the meteorological algorithm is shown in Fig. 7.4. The meteorological algorithm requires only the intensity plane. Hence prior to the detection and inpainting [K. A. Patwardhan and G. Sapiro, 2003, Patwardhan et al., 2005, Gu et al., 2004] of the rain pixels, input RGB frame is converted into the YC_bC_r color space. Chrominance components (C_b and C_r) remain unchanged. In order to detect rain pixels, five consecutive frames have been examined. If there is a relative motion between the camera and the scene then corresponding pixel locations in consecutive frames are obtained by global motion compensation. Once the correspondence of pixels is established in consecutive frames, the change in intensity of the corresponding pixels are examined. According to their intensity changes, possible rain candidates are obtained as given in section 7.3.1. Among possible rain candidates, there are some false positives. In order to refine the rain candidates, area, aspect ratio and orientation of each candidate are matched with the values obtained by the meteorological data. At the end to reduce the visibility of rain, all the potential rain pixels are inpainted with the help of the temporal mean.

7.5   DATABASE

Figure 7.5: Videos used for the simulation (a) "black car02" [K. Garg and S. K. Nayar, 2006], (b) "window' [K. Garg and S. K. Nayar, 2006], (c) "blue car" [S. Starik and M. Werman, 2003], (d) "football" [A. K. Tripathi and S. Mukhopadhyay, 2010], (e) "building" [Barnum et al., 2008].

Simulation is carried out in videos shown in Fig. 7.5. All of the video sequences are reported earlier.

7.6   SIMULATION AND RESULTS

Simulation is performed in a MATLAB 7.0.4 environment. All possible rain candidates can be obtained if consecutive frames are properly aligned. For that purpose the global motion compensation algorithm is applied as explained in the Chapter 6.

The presented meteorological properties-based algorithm requires five consecutive frames which result in high detection rate with low latency. Results of the identification of rain streaks by the meteorological algorithm are shown in Fig. 7.6. Results show that the above mentioned meteorological properties (area, aspect ratio and orientation) of the rain streaks are sufficient for the identification of all potential rain streaks. There is also some miss classification. The reason for the miss classification is that there are intersecting rain streaks producing undesirable shapes that do not match with the properties of the individual rain streak. It is also noted that in sequences taken from movies, the meteorological properties of some raindrops do not fall in the admissible range. The most probable explanation of the failure is that the rain in movies is created using a rain tower, and hence raindrops do not achieve terminal velocity resulting in a shorter streak [N. Brewer and N. Liu, 2008]. The admissible range of area and aspect ratio for identifying a streak may be broadened to identify these artificial raindrops.

We have compared the meteorological algorithm with the prior art algorithms (i.e., Liu et al., temporal and spatiotemporal). These algorithms are selected for their high rain removal and low scene distortions performance. Algorithms are compared in terms of the miss and false detection. Miss detection means that a rain pixel is detected as a non-rain pixel. Low value of miss detection indicates good performance. False detection means that a non-rain pixel is detected as a rain pixel. Low value of false detection indicates good performance. Results of the miss and false detection are shown in Table 7.1. Results are shown for both fixed ('blue car' and "football") and moving ('black car02' and 'window') camera videos. GMC is applied for the videos that are captured by a moving camera irrespective of the rain removal algorithm. The Liu et al. method requires three consecutive frames which give less delay and lot of saving in buffer size, but large miss and false detection degrade the image quality. Temporal and spatiotemporal algorithms are found to be the best two algorithms in the experiments. The meteorological algorithm performs very closely to the temporal and spatiotemporal algorithms in terms of accuracy. Temporal and spatiotemporal algorithms are associated with high latency as fifteen and nine consecutive frames are used by the respective algorithms. The meteorological algorithm has low latency (saving of 45% and 66% buffer size compared to spatiotemporal and temporal algorithms respectively) and removes rain with comparable accuracy. Table 7.1 shows that errors of the said algorithms are small and very close to the temporal and spatiotemporal (2–3 % of the total frame size). For example for video "black car02" error due to temporal, spatiotemporal and meteorological algorithm are 3818, 3756 and 3791 pixels respectively. The numbers of error pixels are 2.14%, 1.98% and 2.00% of the total frame size for the respective algorithms. The meteorological algorithm gives large savings in buffer size and delay with a little sacrifice in the image quality.

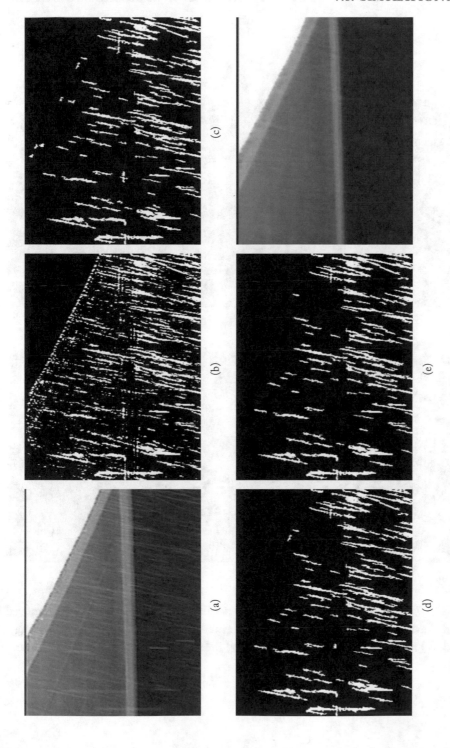

Figure 7.6: (a) A frame from video sequence captured by moving camera, (b) all potential rain streaks (initial rain mask), (c) initial rain mask refined by area property, (d) initial rain mask refined by both area and aspect ratio properties, (e) initial rain mask refined by area, aspect ratio and orientation properties (final rain mask). (f) The frame shown in (a) after rain removed by the meteorological algorithm.

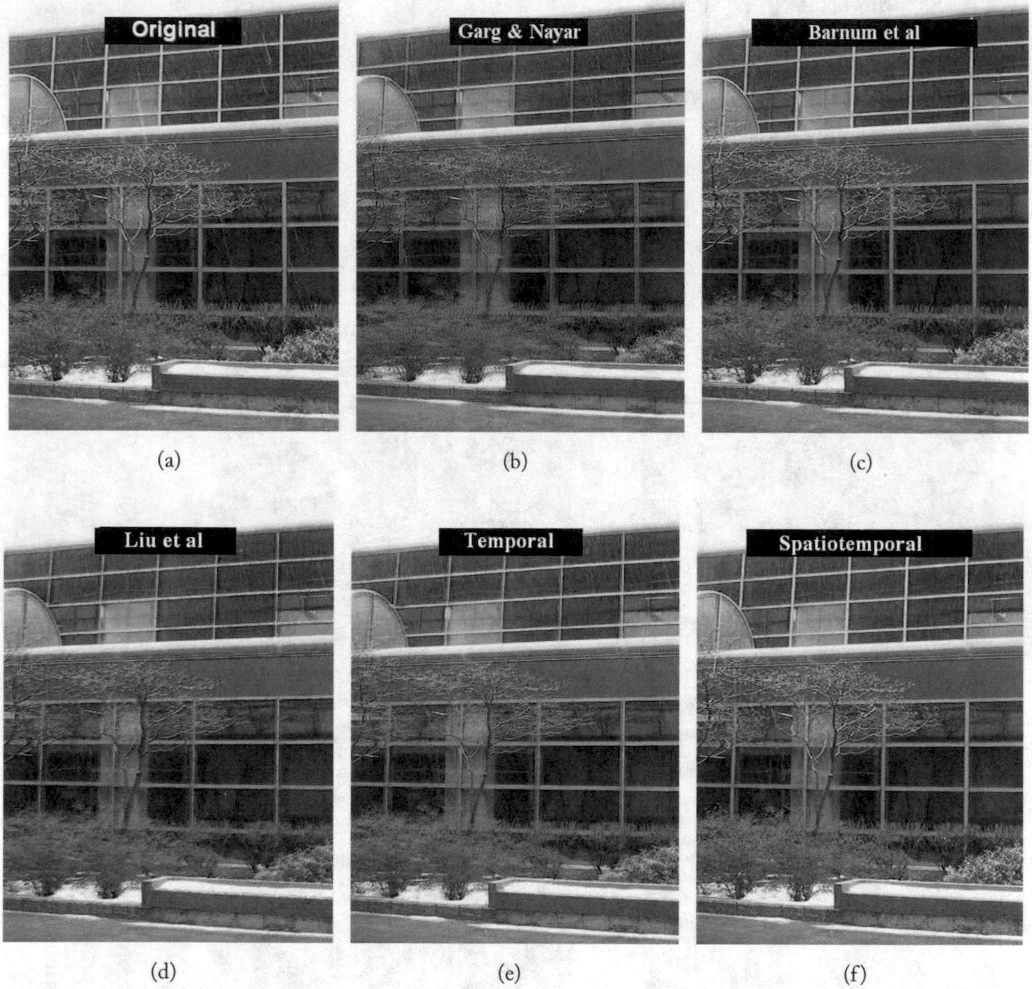

Figure 7.7: (a) The 7th frame of original "building" video. Restored video where rain removed by (b) the Garg and Nayar method, (c) the Barnum et al. method, (d) the Liu et al. method, (e) the temporal method, (f) the spatiotemporal method and (g) the meteorological method. *Continues.*

(g)

Figure 7.7: *Continued.* (g) the meteorological method.

Qualitative analysis for the rain removal algorithms is shown in Fig. 7.7. Results are shown for the "building" video. Results show that the Garg & Nayar method fails to remove rain effectively and produces the unwanted blurring. The Barnum et al. method removes the rain effectively without any unwanted blurring, but fails to detect and remove the rain streak when it is not sharp enough. In the Barnum et al. method rain streaks are considered globally consistent in orientation. Thus, rain streaks located in other orientation (deviation due to turbulent wind flow) are missed. This is in contrast with the meteorological scheme where instead of a particular orientation, a range of orientation is considered. The Liu et al. method removes most of the streaks with fewer distortions but still lots of streaks are missed. The temporal method removes rain effectively but high latency results in blurring. The spatiotemporal method removes the rain with less blurring in comparison with the temporal method. The meteorological method is close to the spatiotemporal method in the removal of rain streaks but associated with less blurring. Thus, the meteorological algorithm outperforms the other existing algorithms in overall performance considering latency, buffer size, accuracy and blurring.

The rain videos ("kgp01" and "kgp02") are captured with a moving camera (SONY® CCD-TRV21E). Results of these videos are shown in Fig. 7.8 and Fig. 7.9. Results show that all the visible rain streaks (in-focus and out-of-focus) are removed perfectly by the meteorological algorithm. More video results can be found at `http://www.ecdept.iitkgp.ernet.in/web/fac ulty/smukho/docs/rain_meteo/rain_meteo.html`.

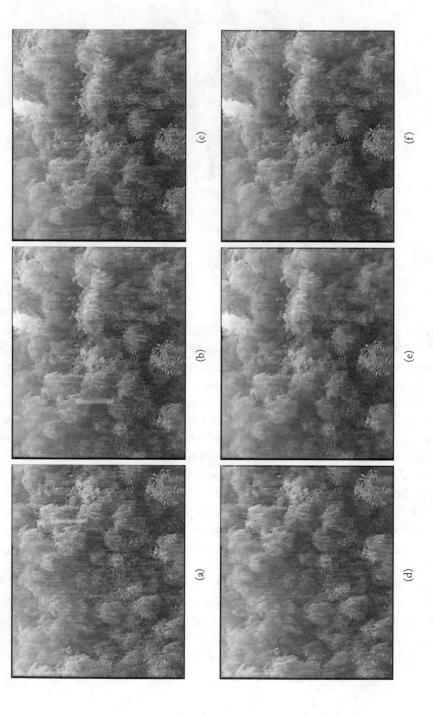

Figure 7.8: (a)–(c) Three frames (62nd, 124th and 140th) of "kgp01" rain video, (d)–(f) removal of rain by the meteorological algorithm in corresponding above frame.

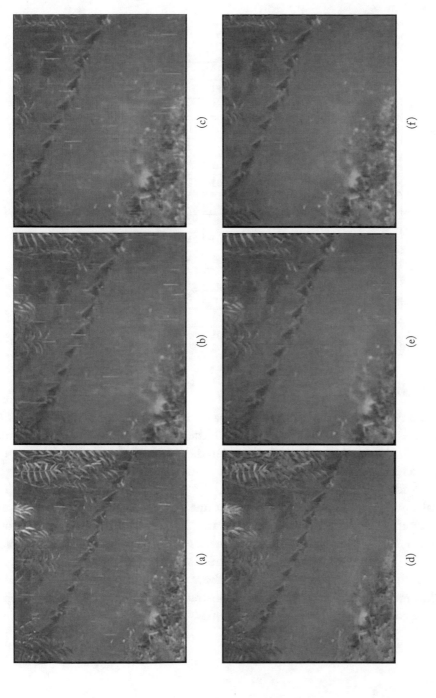

Figure 7.9: (a)–(c) Three frames (124th, 149th and 196th) of "kgp02" rain video, (d)–(f) removal of rain by the meteorological algorithm in corresponding above frame.

Table 7.1: Miss detection (MD) and false detection (FD) of the rain detection algorithms for the different video sequences

Video	Method	MD	FD	Error
"blue car"	Liu et al.	225	1962	2187
	Temporal	47	1822	1869
	Spatiotemporal	62	1824	1886
	Meteorological	78	1793	1871
"football"	Liu et al.	1117	3851	4968
	Temporal	522	2698	3220
	Spatiotemporal	1009	3159	4168
	Meteorological	1211	2472	3683
"black car02"	Liu et al.	756	3479	4235
	Temporal	445	3373	3818
	Spatiotemporal	437	3319	3756
	Meteorological	478	3313	3791
"window"	Liu et al.	679	3297	3976
	Temporal	337	3117	3454
	Spatiotemporal	313	3008	3321
	Meteorological	381	3012	3393

7.7   CONCLUSION

In this chapter, a novel meteorological property-based rain removal algorithm is presented. The said algorithm uses the meteorological properties of the rain. Here relation between the orientation of the rain streak and wind velocity is formulated. Various shape features (viz., aspect ratio) of the rain streaks are also calculated. These shape features help for the discrimination between the rain streaks and moving objects.

It is analyzed that rain gives positive fluctuations in the intensity values, and chrominance values remain unaffected [K. Garg and S. K. Nayar, 2005]. Thus instead of working on all the three color components, the meteorological properties-based algorithm works only on the intensity plane. The use of a single plane reduces the complexity and the execution time of the algorithm. Quantitative and qualitative analysis show that the said meteorological properties-based algorithm removes rain effectively under constrained buffer size and delay in comparison with most of the competing rain removal algorithms.

CHAPTER 8

Conclusion and Scope of Future Work

The first part of this book presents selected works which contributed to the foundation of the video pre/post-processing for combating bad weather due to rain. It has also presented a few latest developments in this arena. The techniques developed using temporal and spatiotemporal properties of rain-affected pixels help to increase the accuracy of detection with a reduced number of frames and do not need any assumption about the shape, size and velocity of the rain drops. This book has also presented temporal and spatiotemporal variance as a reference-free quality metric for measuring the efficacy of the rain removal algorithms. They will be very helpful for assessing rain removal algorithms in real rain as they do not require the ground truth. It is demonstrated that all the rain removal algorithms developed for a fixed camera can be used for moving camera by using global motion estimation. Subsequently a meteorological property-based rain detection algorithm is presented which requires less number of frames and maintains the same accuracy. Here the algorithm makes use of the area, aspect ratio and orientation of the possible rain streaks for the refinement of rain candidates. The developed algorithms require the detection of rain only in intensity plane to reduce the algorithmic complexity and execution time.

In rain detection, future research will focus on the design of the algorithm that requires less number of consecutive frames and optimum number of features, for the discrimination of rain and non-rain pixels with high accuracy. Estimation of features which are robust to the intensity of rain (i.e., heavy, light and moderate rain), background and other environmental effects will help in the detection of rain pixels with higher accuracy. For the classification of rain and non-rain pixels, choice of different features and machine learning algorithms can also affect the results. Use of more accurate and robust estimates of temporal and chromatic properties of rain can increase the accuracy of the rain detection with high perceptual quality. The accuracy of the rain detection suffers due to the presence of moving objects and hence more focus should be given for the detection of moving objects. Care should be taken to ensure parallelism in the algorithm to exploit the architecture of latest processors. The real challenge will be to reduce the number of frames required and the complexity to fit in the power and MIPS budget of hand-held devices. In line with the present focus, there are scopes of development to reduce other unwanted effects of rain, viz., splashes in ground in the "pool" video and the restoration of color of the wet clothes.

Bibliography

R. C. Gonzalez and R. E. Woods, "Digital Image Processing", Addison-Wesley, Reading, Mass., 1992. 61

S. Starik and M. Werman, "Simulation of Rain in Videos", Proceedings of Texture: The 3rd International Workshop on Texture Analysis and Synthesis, France, pp. 95-100, 2003. 3, 8, 15, 34, 35, 65

K. Garg and S. K. Nayar, "Photometric Model of a Rainfrop", Columbia University Technical Report, 2003. 5

K. Garg and S. K. Nayar, "Vision and Rain", International Journal of Computer Vision, Vol. 75, No. 1, pp. 3-27, 2007. DOI: 10.1007/s11263-006-0028-6. 1, 3, 8, 15, 17, 35, 44, 49, 63

K. Garg and S. K. Nayar, "Photorealistic Rendering of Rain Streaks", Proceedings of ACM SIG-GRAPH, Vol. 25, No. 3, pp. 996-1002, 2006. DOI: 10.1145/1141911.1141985. 8, 34, 35, 51, 53, 65

S. G. Narasimhan and S. K. Nayar, "Vision and the Atmosphere", International Journal of Computer Vision, Vol. 48, No. 3, pp. 233-254, 2002. DOI: 10.1023/A:1016328200723. 1, 2

K. Garg and S. K. Nayar, "Detection and Removal of Rain from Videos", IEEE Computer Society Conference on Computer Vision and Pattern Recognition, No. 1, pp. 528-535, 2004. DOI: 10.1109/CVPR.2004.1315077. 13, 15, 16, 17, 21, 35, 44, 49, 63

X. Zhang, H. Li, Y. Qi, W. Leow and T. Ng, "Rain Removal in Video by Combining Temporal and Chromatic Properties", IEEE International Conference on Multimedia and Expo, pp. 461-464, 2006. DOI: 10.1109/ICME.2006.262572. 18, 44, 49, 53, 63

P. Liu, J. Xu, J. Liu and X. Tang, "Pixel Based Temporal Analysis Using Chromatic Property for Removing Rain from Videos", Computer and Information Science, Vol. 2, No. 1, pp. 53-60, 2009. 17, 44, 49, 63

W. J. Park and K. H. Lee, "Rain Removal Using Kalman Filter in Video", International Conference on Smart Manufacturing Application, Korea, pp. 494-497, 2008. DOI: 10.1109/IC-SMA.2008.4505573.

P. Barnum, T. Kanade and S. G. Narasimhan, "Spatio Temporal Frequency Analysis for Removing Rain and Snow from Videos", Workshop on Photometric Analysis For Computer Vision (PACV), In Conjunction with ICCV, 2007. 9, 18, 53

P. Barnum, S. G. Narasimhan and T. Kanade, "Analysis of Rain and Snow in Frequency Space", International Journal of Computer Vision (IJCV), 2009. DOI: 10.1007/s11263-008-0200-2. 9, 18, 65

J. F. Canny, "A Computational Approach to Edge Detection", IEEE Trans. Pattern Anal. Machine Intell., Vol. 8, No. 6, pp. 679-697, 1986. DOI: 10.1109/TPAMI.1986.4767851. 25

A. K. Tripathi and S. Mukhopadhyay, "Rain Rendering in Videos", 4th International Conference on Computer Applications in Electrical Engineering Recent Advances, Roorkee, India, Feb. 2010. 3, 8, 34, 35, 65

A. K. Tripathi and S. Mukhopadhyay, "A probabilistic Approach for Detection and Removal of Rain from Videos", IETE Journal of Research, Vol. 57, No. 1, pp. 82-91, 2011. DOI: 10.4103/0377-2063.78382.

X. Zhao, P. Liu, J. Liu and T. Xianglong, "The Application of Histogram on Rain Detection in Video", Proceedings of the 11th Joint Conference on Information Science, 2008. DOI: 10.1007/s11263-011-0421-7.

K. Garg and S. K. Nayar, "When Does Camera See Rain ?", IEEE International Conference on Computer Vision, No. 2, pp. 1067-1074, 2005. DOI: 10.1109/ICCV.2005.253. 10, 72

J. Bossu, N. Hautiere and J. P. Tarel, "Rain or Snow Detection in Image Sequences through Use of a Histogram of Orientation of Streaks", International Journal of Computer Vision (IJCV), Vol. 93, No. 3, pp.348-367, 2011. DOI: 10.1007/s11263-011-0421-7.

P. Sprent and N. C. Smeeton, *Applied Nonparametric Statistical Methods*, Chapman & Hall/CRC, Third Edition, 2001. 23

R. O. Duda, P.E. Hart and D.G. Stork: *Pattern Classification*, John Wiley & Sons, Second Edition, 2001. 26

N. Brewer and N. Liu, "Using the Shape Characteristics of Rain to Identify and Remove Rain from Video", Proceedings of the 2008 Joint IAPR International Workshop on Structural, Syntactic, and Statistical Pattern Recognition, pp. 451-458, 2008. DOI: 10.1007/978-3-540-89689-0_49. 49, 63, 66

M. F. Subhani, J. P. Oakley, "Low Latency Mitigation of Rain Induced Noise in Images", 5th European Conference on Visual Media Production (CVMP), pp. 1-4, 2008.

K. A. Patwardhan and G. Sapiro, "Projection Based Image and Video Inpainting Using Wavelets", International Conference on Image Procesing, Vol. 1, pp. 857-860, 2003. DOI: 10.1109/ICIP.2003.1247098. 13, 25, 64

K. A. Patwardhan, G. Sapiro and M. Bertalmio, "Video Inpainting of Occluding and Occluded objects", International Conference on Image Procesing, Vol. 2, pp. 69-72, 2005. DOI: 10.1109/ICIP.2005.1529993. 13, 25, 64

J. Gu, S. Peng and X. Wang, "Digital Image Inpainting Using Monte Carlo Method", International Conference on Image Procesing, Vol. 2, pp. 961 - 964, 2004. DOI: 10.1109/ICIP.2004.1419460. 13, 25, 64

A. M. Tekalp, *Digital Video Processing*, First Edition, Prentice Hall of India, 1995. 49

D. Xu, J. Liu, X. Li, Z. Liu and X. Tang, "Insignificant Shadow Detection for Video Segmentation", IEEE Transaction on Circuits and Systems for Video Technology, Vol. 15, No. 8, pp. 1058-1064, 2005. DOI: 10.1109/TCSVT.2005.852402. 25

P. Liu, J. Xu, J. Liu and X. Tang, "A Rain Removal Method Using Chromatic Property for Image Sequence", Proceedings of the 11th Joint Conference on Information Science, Dec. 2008. 17

T. Chen, "Video Stabilization Algorithm Using a Block-Based Parametric Motion Model", Mater's thesis, Stanford University, winter 2000. 49, 55

Y. Su, M. T. Sun and V. Hsu, "Global Motion Estimation from Coarsely Sampled Motion Vector Field and the Applications", IEEE Transaction on Circuits and Systems for Video Technology, Vol. 15, No. 2, pp. 232-242, Feb. 2005. 49

I. Markovsky and S. V. Huffel, "Overview of Total Least-Squares Methods", Signal Process., Vol. 87, No. 10, pp. 2283-2302, 2007. DOI: 10.1016/j.sigpro.2007.04.004. 50

Z. Feng, M. Tang, J. Dong and S. Chou, "Real-Time Rain Simulation", LNCS 3865, pp. 626-635, 2006. DOI: 10.1109/CAD-CG.2005.67. 57

G. K. Batchelor, *An Introduction to Fluid Dynamics*, Cambridge University Press, 2000. DOI: 10.1017/CBO9780511800955. 57, 58, 59

G. K. Siogkas and E. S. Dermatas, "Detection, Tracking and Classification of Road Signs in Adverse Conditions", IEEE MELECON, Spain, pp. 537-540, 2006. DOI: 10.1109/MELCON.2006.1653157. 1

K. V. Beard and C. Chuang, "A New Model for the Equilibrium Shape of Raindrops", Journal of the Atmospheric Sciences, Vol. 44, No. 11, pp. 1509-1524 ,1987. DOI: 10.1175/1520-0469(1987)044%3C1509:ANMFTE%3E2.0.CO;2. 3, 4

J. S. Marshall and W. M. K. Palmer, "The Distribution of Raindrops with Sizes", Journal of Meterology, Vol.5, pp. 165-166, 1948. DOI: 10.1175/1520-0469(1948)005%3C0165:TDORWS%3E2.0.CO;2. 3, 5

R. Gunn and G. D. Kinzer, "Terminal Velocity of Fall for Water Droplet in Stagnant Air", Journal of Meteorology, Vol. 6, pp. 243-248, 1949. DOI: 10.1175/1520-0469(1949)006%3C0243:TTVOFF%3E2.0.CO;2. 3, 5, 59, 60

G. B. Foote and P. S. DuToit, "Terminal Velocity of Raindrops Aloft", Journal of Applied Meteorology. Vol. 8, pp. 249-253, 1969. DOI: 10.1175/1520-0450(1969)008%3C0249:TVORA%3E2.0.CO;2. 5

U.S. Department of Transportation Federal Highway Administration (http://ops.fhwa.dot.gov/Weather/) 1

M. Schonhuber, H.E. Urban, J.P. Baptista, W.L. Randeu, and W. Riedl, 1994. "Measurements of precipitation characteristics of new disdrometer". In Proceedings of Atmospheric Physics and Dynamics in the Analysis and Prognosis of Precipitation Fields. 3

Authors' Biographies

SUDIPTA MUKHOPADHYAY

Sudipta Mukhopadhyay is currently Associate Professor in the Electrical and Electrical Communication Engineering, IIT Kharagpur. He received his B.E. degree from Jadavpur University, Kolkata, in 1988. He received his M.Tech. and Ph.D. degrees from IIT Kanpur in 1991 and 1996 respectively. He has served several companies including TCS, Silicon Automation Systems, GE India Technology Centre and Philips Medical Systems before joining IIT Kharagpur in 2005 as Assistant Professor of Electrical and Electrical Communication Engineering, IIT Kharagpur. In 2013 he become Associate Professor in the same department. He has authored or co-authored more than 70 publications in the field of signal and image processing. He has filed seven patents while working in industry and continued the trend after joining academia. He is a senior member of the Institute of Electrical and Electronics Engineers (IEEE), Member of SPIE and corresponding member of Radiological Society of North America (RSNA). He has done many applied projects sponsored by DIT, Intel and GE Medical Systems IT, USA. He is also founder and director of Perceptivo Imaging Technologies Private Ltd., a company under the guidance of S.T.E.P. IIT Kharagpur. The company specializes in developing innovative software for signal and image processing.

ABHISHEK TRIPATHI

Abhishek Tripathi is currently working as Senior Engineer at Uurmi Systems Pvt. Ltd., Hyderabad, India. He received his Ph.D. degree from Indian Institute of Technology Kharagpur, India, in 2012. He received the M.Tech. degree from National Institute of Technology, Kurukshetra, India, in 2008. He received the B.Tech. degree from Uttar Pradesh Technical University, Lucknow, India, in 2006. His research interests include computer vision, image-based rendering, nonlinear image processing, physics-based vision, video post processing, recognition, machine learning and medical imaging.

Printed in the United States
by Baker & Taylor Publisher Services